HANDBOOK OF MODEL
ROCKETRY

HANDBOOK OF MODEL ROCKETRY

by G. Harry Stine

ARCO PUBLISHING, INC.
NEW YORK

First Arco edition, first printing

Published by Arco Publishing, Inc.
215 Park Avenue South, New York, N.Y. 10003

Copyright © 1965, 1967, 1970, 1976, 1983 by G. Harry Stine.

Illustrations copyright © 1976, 1982 by G. Harry Stine.

Originally published by Follett Publishing Company, Chicago.

Library of Congress Cataloging in Publication Data

Stine, G. Harry (George Harry), 1928–

  The handbook of model rocketry.

  Bibliography: p. 333.
  Includes indexes.
  1. Rockets (Aeronautics)—Models. I. Title.
TL844.S77   1983   629.46′52′0228   82-8913
ISBN   0-668-05358-5   (Reference Text) AACR2
ISBN 0–668–05360–7 (Paper edition)

Printed in the United States of America

10   9   8   7   6   5   4   3   2   1

About the Author

G. Harry Stine is probably the person most responsible for starting model rocketry and guiding it through its early years as it grew into a highly respected worldwide aerospace hobby and sport. In 1957, he founded the National Association of Rocketry (NAR) and started the first model rocket company, Model Missiles, Inc. An engineer, science writer, and science fiction author, he has written more than two dozen books since 1950. His many science fiction stories have appeared under his pen name, "Lee Correy," but he uses his real name for nonfiction writing. This book, "The Handbook of Model Rocketry," has been cited as one of his most influential works and is in its fifth edition as the "bible" and official reference book of the entire hobby; its success has prompted many other books on model rocketry here and around the world, but it remains the updated original and most used book about model rocketry.

Stine not only founded the NAR, but served for ten years as its president; he is now an Honorary Member and an Honorary Trustee of the NAR. For eleven years he served as President of the Space Model Subcommittee of the Federation Aeronautique Internationale (FAI) in Paris. He represents the NAR as the Chairman of the Committee on Pyrotechnics of the National Fire Protection Association. He has been United States Senior National Champion Model Rocketeer four times and has guided his two daughters and son to United States Junior Championships. His daughter, Ellie, won a bronze medal in the First World Championships for Space Models held in Vrsac, Yugoslavia, in 1972, and he served as Chairman of the International Jury for that Olympic-level competition as well as for the World Championships in Dubnica-nad-Vahom, Czechoslovakia, in 1974. He is NAR #2 and holds United States Sporting Aviation License N-002.

His list of honors and memberships is long: Fellow of the Explorers Club, Fellow of the British Interplanetary Society, Associate Fellow of the American Institute of Aeronautics and Astronautics, Scientific Leader Member of the Academy of Model Aeronautics, Member of the New York Academy of Sciences, and Member of the Aircraft Owners and Pilots Association. He is listed in Who's Who In America. In 1969, he received a special award from the Hobby Industry Association for his part in establishing the model rocket industry and a silver medal as one of fifty United States space pioneers honored by the United States Army Association.

He's also an instrument rated private pilot who owns his own airplane, a Cherokee, his "manned boost-glider."

He lives in Phoenix, Arizona, with his wife, three Golden Retrievers (his recovery crew), and a cat with 24 toes. He does all his writing on a minicomputer that is also programmed to do many of the calculations discussed in this book.

Thousands of model rocketeers know him as "The Old Rocketeer."

to
William S. Roe
Orville H. Carlisle
and a couple of thousand
young and old model rocketeers

Contents

Preface

Every author has a book or story that he or she has always wanted to write. In my case, this is it. I labor under the delusion that it might be more important than all the fiction I've done and of broader consequence than much of my nonfiction.

I hope it may save the hands, eyes, and lives of countless youngsters who might never have learned about model rocketry without it. I also hope it may set many young people on their course toward becoming astronauts, engineers, technicians, and other kinds of scientists. Finally, I hope it may serve as a guidepost to many people, young and old, who are interested in rockets.

Model rocketry is my hobby. I have given much time and effort to it and I have gained rewards far more valuable than mere money from it. The basic motive for my involvement in model rocketry stems from my youth, when many scientists, engineers, teachers, and other adults freely gave me advice, guidance, help, and the means to do things.

Once, I asked one of these men what I could do to repay him. "Do the same for others when you grow up," he told me. In the Space Age hobby of model rocketry, I have found a way to do this.

—G. Harry Stine

Introduction to the Fifth Edition

When this book was originally written in 1963, it was the first comprehensive model rocketry manual. It was intended to serve as a complete handbook for the entire hobby. I tried to cover every aspect of the growing young hobby to keep the beginner from "reinventing the wheel"—i.e., making the same mistakes others had made and learned from. I also wanted to lead the modeler into the interesting, advanced aspects of model rocketry.

The four previous editions brought out by another publisher in 1965, 1967, 1970, and 1976 more than fulfilled these aims. The *Handbook* has had a major impact on the hobby of model rocketry. I know the full extent of this impact from my conversations and correspondence with thousands of model rocketeers from all over the world.

As with any book based on technology—and model rocketry is a technology in miniature—the *Handbook* has gone thoroughly out of date since the fourth edition was published in 1976. And, because model rocketry has grown as a technology and developed highly specialized aspects, this *Handbook* can no longer deal in detail with *all* aspects of model rocketry. The best I can now do is to present most of the basics that can serve as a foundation for the more advanced areas, and then merely inform the reader of the rudiments of these more advanced areas. This will, I hope, lead the reader into advanced topics beyond the scope of this book. The technical state of the art has now progressed to the point where model rocketry is routinely used not only at the elementary and secondary school levels, but also at the university undergraduate and graduate levels at such colleges and universities as the Massachusetts Institute of Technology, Ohio State University, and the United States Air Force Academy. Many colleges and universities now host annual model rocketry conventions, and the United States National Model Rocket Championships (NARAM) has been held many times at such places in addition to NASA, Army, and Air Force installations.

The *Handbook* has never been a theoretical work divorced from the realities of the workshop and flying field. I've used previous editions as texts for comprehensive model rocketry training courses since 1965 in New Canaan, Connecticut, and Phoenix, Arizona. The layout, format, approach, and content are based upon positive, in-the-field feedback from model rocketeers of all ages who had fun, learned something about science and technology, and unknowingly contributed to this *Handbook* in its evolutionary development through the years.

All of them have learned, as you and thousands of others have and will, that model rocketry is like wrestling with a bear: It's awfully hard to disengage yourself from it. No other hobby can claim to encompass as many different areas of modern science and technology. And no other aerospace hobby is as safe, as easy, and as inexpensive to participate in.

Model rocketry—or "space modeling," as it is known outside the United States—has leaped oceans and political ideologies to become a medium of communication between people all over planet earth who got involved in model rocketry with starlight in their eyes. In the process, they discovered what science and technology are all about and what will be required of them of they wish to go to the stars.

People will go to the stars someday, and model rocketry may help pave the way.

Phoenix, Arizona   —G. Harry Stine

1

This Is Model Rocketry

Model rocketry has been called miniature astronautics, a technology in miniature, a hobby, a sport, a technological recreation, an educational tool— and it *is* all of these things. It is a safe, enjoyable, and highly respected pastime that now boasts the enthusiastic participation of millions of people, young and old, in the United States, Canada, Great Britain, Australia, the Netherlands, the Federal Republic of Germany, Spain, Switzerland, the German Democratic Republic, Czechoslovakia, Poland, Bulgaria, Rumania, Yugoslavia, and the Union of Soviet Socialist Republics.

Model rocketry started in the United States in 1957. Its beginnings have been carefully and thoroughly documented. Models, correspondence, drawings, publications, and other artifacts and documents concerning the precise details of the early years have been donated to the National Air and Space Museum of the Smithsonian Institution, where this extensive collection can be properly preserved for posterity. The sense of history, of doing something new and unique in the world, ran high among those of us who participated in those early years. As a result, an enormous collection of significant things was saved.

Model rocketry resulted from a timely combination of model aeronautics, the ancient art of pyrotechnics, and modern rocket technology. Although all of these elements had existed separately for over a decade before 1957, it fell to two men to combine them successfully into a Space Age hobby.

The first model rockets were built and flown by Orville H. Carlisle, the owner of a shoe store in Norfolk, Nebraska. With his brother's help, Carlisle designed the first model rockets, bringing together model aeronautics and pyrotechnics. Early in 1957, I added the elements of professional rocket technology, and today's model rocketry was born.

What is model rocketry? What makes a model rocket so safe, so inexpensive, so easy to build, and so much fun to fly that millions of people have successfully launched them since the first kits and motors appeared in April 1958? Why has model rocketry brought the Space Age to Main Street, directly involving more people in rocketry than have ever watched a Space Shuttle launch from Cape Canaveral or Vandenberg Air Force Base?

1

Figure 1-1 Model rocketry is a national and international sport and hobby. Here a model rocket lifts off from the Mall in Washington, DC, with the nation's Capitol in the background. The occasion was the annual Mall Demonstration of the National Association of Rocketry sponsored by the National Air and Space Museum of the Smithsonian Institution.

Figure 1-2 Model rocketry was started by the author (left) and Orville H. Carlisle (right), founders of the National Association of Rocketry (NAR) shown here comparing model rockets at the Ninth Annual Model Rocket Championships.

A model rocket is an aerospace model, a miniature version of a real space vehicle, with *all* of the following characteristics:

1. It is made of paper, balsa wood, plastic, cardboard, and other nonmetallic materials without any metals used as structural parts.
2. It weighs less than 16 ounces and carries less than 4 ounces of rocket propellant, in accordance with the regulations of the Federal Aviation Administration.
3. It uses factory-made, preloaded, nonmetallic, expendable, solid-propellant rocket reaction motors that are replaced after each flight. This eliminates any hazard of mixing or handling dangerous propellant chemicals.
4. Its solid-propellant rocket motor is ignited electrically from a distance of 15 feet or more, using a battery and an electrical launch controller with built-in safety features that allow the motor to be ignited on countdown only when desired.
5. It contains a recovery device (or devices) to lower it gently and safely back to the ground; thus it can be flown again and again by installing a new solid-propellant rocket motor and repacking the recovery device.

Figure 1-3 This typical model rocket beginner's kit, the Estes Alpha III, has paper and plastic parts, parachute recovery, and molded plastic tail assembly. The finished model stands ready for flight.

A model rocket is that simple, yet it changed the nature of nonprofessional rocketry. Before 1958, nonprofessional rocketry was so dangerous that it was banned by law in most states. The American Rocket Society (now the American Institute of Aeronautics and Astronautics) estimated that one out of seven nonprofessional rocketeers would be injured or killed in their avocation. Today, model rocketry is so ubiquitous, so available, and so safe that one can launch almost anywhere with complete freedom—and in safety, provided one follows the simple safety rules such as the Safety Code of the National Association of Rocketry (NAR) and the Hobby Industry Association (HIA). Because of its outstanding safety record, model rocketry enjoys the favor of the Federal Aviation Administration (FAA), and the prestigious and

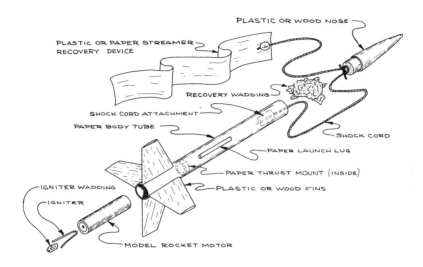

Figure 1-4 A sketch of a typical model rocket showing its various parts. There are many variations on the size and shape of many of the parts, but all model rockets are basically similar.

safety-conscious National Fire Protection Association (NFPA). Model rocketry products have been carefully evaluated by the Department of Transportation, the Bureau of Explosives of the Association of American Railroads, the Air Line Pilots Association, the Food and Drug Administration, and the Consumer Product Safety Commission. Model rocketry enjoys the enthusiastic support of the National Aeronautics and Space Administration (NASA), the United States Air Force, the United States Army, the United States Navy, the Boy Scouts of America, the 4-H Clubs of America, the YMCA and the YMHA, the Civil Air Patrol, and the American Society for Aerospace Education. Even the highly professional American Institute of Aeronautics and Astronautics reversed its early stand against nonprofessional rocketry.

This confidence is deserved. As of 1981, twenty-three years after the first model rocket kits and motors became commercially available, nearly 200 million model rockets have been flown successfully and safely in the United States alone. The hobby is far safer than swimming, boating, baseball, football, and cycling. Since 1964, the members of the National Association of Rocketry (NAR) have been covered by a major liability insurance policy underwritten by a large domestic insurance firm.

As of this writing, all but three states (Massachusetts, Rhode Island, and New Jersey) have adopted the "Code for Unmanned Rockets" (NFPA No. 1122) of the National Fire Protection Association, which is a permissive code requiring extensive product testing and quality control by the model rocket manufacturers and no permits for sale, purchase, possession, or use by merchants or the public.

So don't let anybody tell you that model rocketry is hazardous to people or property. The facts prove otherwise.

Naturally, it's possible to get hurt in model rocketry if you're stupid, don't follow the NAR/HIA Safety Code, and don't read and follow instructions. It's possible to get hurt or to have an accident with *any* sort of technology, new or old. Some people even manage to get hurt while collecting stamps.

Model rocketry's excellent safety record is due primarily to the non-metallic, lightweight construction of model rocket airframes, the preloaded model rocket motor, and the voluntary limitation the hobby has upon maximum weight.

It's not obvious at first that you can learn anything about rockets if you use a ready-made rocket motor. But the science of astronautics is more than a matter of rocket motors and propulsion. The business of making a rocket motor of *any* type or size is a very complicated, expensive, dangerous, delicate affair that *must not* be attempted by anyone with less than many thousands of dollars for the proper equipment, an advanced college degree in chemistry, several years' experience in handling explosives, several acres of land providing a safe place to work, and a very large life-insurance policy. (Three professional model rocket motor technicians have been killed while making model rocket motors under the most carefully controlled conditions with all of the safety equipment available. The model rocket motor manufacturing companies therefore take grave risks and assume the awful hazards of rocket motor manufacture so that model rocketeers can enjoy their hobby in safety.)

Model rocketry is very similar to model aviation regarding motor construction. If you wanted to build a powered model airplane, it would be very expensive and time-consuming if you had to build the *entire* model, including the little gasoline engine. You have more fun and more flying time if you build only the model airplane and purchase the motor ready-to-run. Then you know you have a motor that will work and—if you're a careful modeler and flyer—an airplane that will usually get high enough to crash. . . .

The same holds true for model rocketry. By properly using a factory-made, preloaded model rocket motor and building a successful model rocket airframe around it, you can have a flying model rocket with an excellent chance of flying successfully and safely. Furthermore, you will eventually learn about rocket thrust, duration, total impulse, specific impulse, grain configuration, thrust–time performance, and other rocket motor facts. Since your model is a free body in space even when it's in the air, its actions in flight are quite different from those on the ground on wheels or skids. Subtract the known effects of the surrounding air and you will be able to understand how a rocket behaves in the vacuum of space. Flight performance can tell you a great deal about how things move in the universe. And even the fact that your model rocket always flies within the earth's atmosphere will introduce you to the fascinating mysteries of aerodynamics, weight and balance, stability, and drag. In finding out more about why your model rockets fly as they do (and sometimes why they don't fly as predicted), you can begin to delve into optics, structures, dynamics, electronics, meteorology, materials science, and many other modern technologies including the mathematical tools that make them useful to mankind. Model rocket design and flight performance are also excellent areas where computer analysis can be brought into play, and many good BASIC computer programs exist that permit you to use simple minicomputers or pocket calculators to determine model rocket design and flight factors.

A model rocket is wonderfully simple to build. You can use common hobby tools of the sort used to build model airplanes, model cars, and model boats. Even the materials employed in these other model hobbies are used in model rocketry. Many people think that a rocket has to be made out of steel or other metals, but that isn't so. Why make a rocket out of metal when you can make one cheaply out of paper, plastic, and wood? Why spend a lot of money to buy a metal-welding outfit when a fifty-cent tube of model airplane glue will do the same job?

Besides ease of construction, there are other reasons for using nonmetallic materials. High strength and light weight have always been prime design goals for flying devices of any type. Paper and balsa wood meet these requirements well. Even some of the new composite plastics are beginning to find their way into model rocketry. All of these nonmetallic materials can be used in a model rocket. For example, a model rocket motor with its paper casing is barely warm to the touch immediately after operating. This is because paper is a very good insulator and does not readily conduct heat. Although much lighter in weight than metal, paper and balsa wood are surprisingly strong. In fact, balsa wood has a higher ratio of strength to weight than carbon steel.

In addition, nonmetallic materials are much safer should something go wrong during the flight of a model rocket. A model rocket made of paper, balsa, and plastic literally disintegrates if it happens to hit something. It absorbs the impact by destroying itself. Model rockets have been deliberately launched point-blank into sheets of window glass; these experiments completely destroyed the models but didn't even scratch the glass.

All model rockets have recovery devices. One might ask, "Why bother about a recovery device?" Answer: because you want your model rocket to come down and land gently in a condition to fly again, just like the NASA

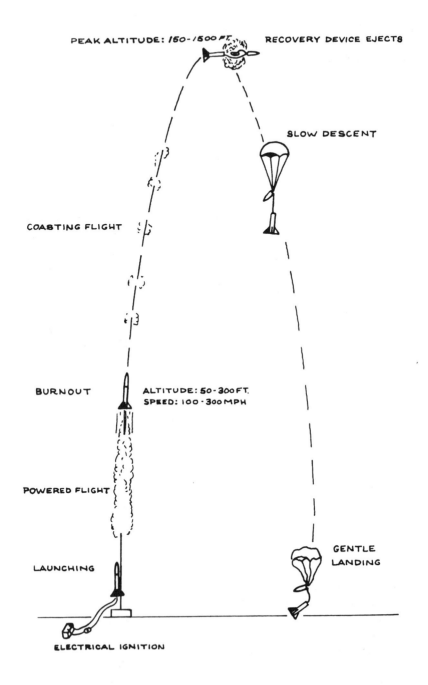

PEAK ALTITUDE: *150-1500 FT.*   RECOVERY DEVICE EJECTS

SLOW DESCENT

COASTING FLIGHT

BURNOUT   ALTITUDE: 50-300 FT.
SPEED: 100-300 MPH

POWERED FLIGHT

GENTLE
LANDING

LAUNCHING

ELECTRICAL IGNITION

Figure 1-5 A typical flight of a single-staged model rocket with parachute recovery.

Figure 1-6 Recovery! At the end of its flight a model rocket lands so gently with its recovery device that it can actually be caught bare-handed in midair. Install a new model rocket motor and repack the recovery device, and the model is ready for another flight. (Centuri Engineering Company)

Space Shuttle. After you've spent some money and taken some time to build a model rocket, you'll want to get more than a single flight out of it. Recovery devices are very easy to install, and they work with exceedingly high reliability. Furthermore, they help make model rocketry a safe activity. Even a 1-ounce model rocket plummeting back to the ground from an altitude of 500 feet can hurt if it hits you. A recovery device to retard the speed of its descent eliminates this potential hazard. The mandatory use of recovery devices is one of the reasons why model rocketry has been such a safe activity.

But why build model rockets so small? Why not build them big so that they're impressive, go several miles high, and carry lots of payload like the "big ones" at Cape Canaveral, Vandenberg Air Force Base, and Wallops Flight Research Center? There are several reasons. Model rockets are built small for the sake of cost, simplicity, and safety. The price of a model rocket goes up even faster than a rocket itself as the size increases; you can look at any model rocket catalog for proof of this. Big birds get to be very expensive. And as model rockets become bigger, they become more difficult and demanding to build. Furthermore, as model rockets become bigger, they become less safe. Large model rocket motors are required for large model rockets, and large model rocket motors are expensive. To promote safety and because model rockets really don't *need* very much thrust, duration, or total impulse to achieve impressive performances, model rocket motors have been limited in power. A model rocketeer doesn't use brute-force methods; he does careful work in design and construction to get the ultimate performance from a given amount of limited rocket motor power.

Model rockets don't have to be big to carry interesting payloads. Using ingenuity and technical know-how, model rocketeers manage successfully to fly and recover such payloads as fresh hen's eggs, radio transmitters, still cameras, and even movie cameras, all within the limitation of a maximum weight of 1 pound.

But why limit the rocket power? Doesn't this prevent a model rocketeer from launching models to really impressive high altitudes?

Yes, it does. And what are you going to do once the model rocket goes out of sight, a feat easily accomplished even with limited motor power? Small models go out of sight at altitudes between 1,000 and 1,500 feet; larger models become invisible above 2,500 feet—less if the visibility is poor. Of course, it's possible to put a model rocket much higher than this. But why do it? Once it's out of sight, you need radio tracking equipment, tracking telescopes, and large tracts of land. You might as well go to work for NASA in the first place.

Today, *anybody* can put a large model rocket motor in a very small, highly streamlined model rocket and launch it straight up and out of sight. It doesn't prove that the person who did it is a genius, somebody special, a technology wizard, or anything of the sort. It does, however, indicate that the person is a "technology twit"—one who doesn't care how science and technology are used, which is a matter of considerable concern these days.

Sheer altitude isn't the only goal in model rocketry. There are a lot of things to do that are more fun, more challenging, and provide a real sense of accomplishment.

To begin with, you'll work hard getting your first model rocket together and off the ground for its first successful flight. You can do it—*if you read and follow all instructions carefully!*

If at first you don't succeed, try reading the instructions.

After your first launch or two, you'll be striving for reliable flights, reliable ignition, straight flight paths, full recovery deployment, recovery in the same county, getting the model out of trees, and getting it ready for more flights. You'll progress to more difficult models, to motors of higher power, to multistaged models, to glide-recovery models, and to payload-carrying models. You'll be able to try your hand at scale model rockets. Or perhaps you'll design your own model and experience the thrill of seeing it work exactly as you predicted. Maybe you'll enter contests and, after learning all the ins and outs of model rocket competition, start to win ribbons, trophies, and prizes.

As you grow more deeply involved in model rocketry, you'll find yourself studying a wide variety of subjects in order to understand what your model rockets are doing and how to improve them. You'll *never* have to worry about a science fair project if you're a student rocketeer.

When a model rocketeer talks with a professional rocketeer, they speak the same language. They can talk about specific impulse, ballistics, data reduction, static testing, spin stability, recovery parachute deployment, lift-to-drag ratios, and hundreds of other topics. And, surprisingly, they have a great deal of mutual respect for one another. Many professional rocket engineers are also model rocketeers. And now, after a quarter of a century, many young people who started out as model rocketeers have become professionals in the aerospace sciences and engineering fields. I can hardly go to a NASA center or an aerospace company without meeting somebody whom I last saw years before on a model rocket range, as a teenage student taking the first steps toward the stars.

This is because model rocketry is indeed a technology in miniature and was deliberately designed that way. It recognizes that there are many interesting aspects to aerospace technology other than the propulsion system. To model rocketeers, a model rocket motor is nothing more than a prime mover, a realistic device that produces the thrust force to lift their models into the sky.

Model rocketry has become many things to many people. It can be a means to learn something about the universe. It can be a way to satisfy the competitive spirit that says, "My model is better than your model!" It can be a way to learn and a way to teach others. For many people, it is an enjoyable

Figure 1-7 Model rocketry is international, having spread from the United States around the world. Here Juliuz Jaronzyk of the Polish team puts his scale model French "Diamant" satellite launcher on the pad at the First World Championships for Space Models held in Vrsac, Yugoslavia.

TABLE 1
NAR-HIA Model Rocket
Safety Code

1. *Construction:* My model rockets will be made of lightweight materials such as paper, wood, plastic, and rubber, without any metal as structural parts.

2. *Engines:* I will use only preloaded factory-made NAR Safety-Certified model rocket engines in the manner recommended by the manufacturer. I will not change in any way nor attempt to reload these engines.

3. *Recovery:* I will always use a recovery system in my model rockets that will return them safely to the ground so that they may be flown again.

4. *Weight Limits:* My model rocket will weigh no more than 453 grams (16 ounces) at lift-off, and the engines will contain no more than 113 grams (4 ounces) of propellant.

5. *Stability:* I will check the stability of my model rockets before their first flight, except when launching models of already proven stability.

6. *Launching System:* The system I use to launch my model rockets must be remotely controlled and electrically operated, and will contain a switch that will return to "off" when released. I will remain at least 15 feet away from any rocket that is being launched.

7. *Launch Safety:* I will not let anyone approach a model rocket on a launcher until I have made sure that either the safety interlock key has been removed or that the battery has been disconnected from my launcher.

8. *Flying Conditions:* I will not launch my model rocket in high winds, near buildings, power lines, tall trees, low flying aircraft, or under any conditions which might be dangerous to people or property.

9. *Launch Area:* My model rockets will always be launched from a cleared area, free of any flammable materials, and I will only use nonflammable recovery wadding in my rockets.

10. *Jet Deflector:* My launcher will have a jet deflector device to prevent the engine exhaust from hitting the ground directly.

11. *Launch Rod:* To prevent accidental eye injury I will always place the launcher so that the end of the rod is above eye level or cap the end of the rod with my hand when approaching it. I will never place my head or body over the launching rod. When my launcher is not in use I will always store it so that the launch rod is NOT in an upright position.

12. *Power Lines:* I will never attempt to recover my rocket from a power line or other dangerous places.

13. *Launch Targets and Angle:* I will not launch rockets so their flight path will carry them against targets on the ground and will never use an explosive warhead nor a payload that is intended to be flammable. My launching device will always be pointed within 30 degrees of vertical.

14. *Prelaunch Test:* When conducting research activities with unproven designs or methods, I will, when possible, determine their reliability through prelaunch tests. I will conduct launchings of unproven designs in complete isolation.

recreation or hobby that combines the individualistic craftsmanship of the home workshop and the happy socializing with others on the flying field, in the sunshine and fresh air. It can be a way for young and old to get together in an activity that interests both.

Model rocketry combines modern science and technology, craftsmanship and shop practice, individual creativity and group cooperation, and the pursuit of excellence in a healthy outdoor activity. Sportsmanship, craftsmanship, self-reliance, discipline, and pragmatic thinking are only a few of the things that can be learned in model rocketry.

But, mostly, model rocketry is fun—as long as you follow the Safety Code.

You'll never run out of things to do in model rocketry; it's an endless countdown, a hobby that you can grow up with and stay with for years because it offers a never-ending challenge to build better model rockets.

Congratulations on getting hooked on the best aerospace hobby this side of Alpha Centauri.

2

Getting Started

You can get started in model rocketry for between ten and twenty dollars, depending upon the model and equipment you select. This will set you up with a launch pad, an electric launch-controller system, a launching battery, a simple beginner's model rocket kit, and a few model rocket motors.

All model rocket manufacturers (see Appendix I for their names and addresses, correct as of this writing) make and sell starter sets that include all of the above equipment. However, some of the starter sets do not include the battery, which must then be purchased separately. You can purchase all of this equipment separately if you wish, but the prepackaged starter set will save you time and money. In any event, you will continue to use the launch pad, the electric launch controller, and the launching battery in all your model rocket flying activities for a long time. So this big initial cost hits you only once. If you intend to pursue the hobby seriously, you should get the best launch pad and controller that you can afford and obtain a large, rechargeable battery such as a "gel cell" with its recharging device.

If you are handy with tools, you can make your own launch pad and electric launch controller. You'll find the details about this in the chapter about launching.

Don't try to fly a model rocket without a launch pad!

Don't attempt to use any ignition system other than electric ignition with a battery and launch controller in accordance with the instructions of the manufacturer of the model rocket motor you're using.

A quarter of a century of experience in model rocketry and more than fifty years of professional rocket activity have proven the validity of these statements and the safety of these two necessary items. The few accidents that have occurred in model rocketry have been caused by attempts to take shortcuts in the requirements and safety rules for launch pads and electric ignition. It isn't worth it to try to shortcut experience. If you're going to do something, do it right or don't do it at all. If you're stupid enough to think you're smart enough to ignore all the technical experience and know-how and do it *your way*, no model rocketeer will shed a single tear for you when you have your inevitable accident. In fact, they'll be pretty angry with you for being a bad example when they work so hard to do it the right way, the safe way.

Figure 2-1 This typical starter set has nearly everything you'll need to get started in model rocketry—a model rocket kit, a launch pad, an electric launch controller with Size AA penlight cells inside, recovery wadding, three model rocket motors, and complete instructions. You need supply only simple tools and the ability to follow instructions. (Estes Industries, Inc.)

If you're a youngster, you'll probably need some help from your dad or another adult to assemble the launch pad and controller properly. In general, people less than ten years old don't have the manual dexterity to build model rockets and launch equipment without help. *Don't be stupid and fail to ask for help* if you get into a bind! The instructions that come with model rocket kits, motors, and equipment are probably the best instructions of any activity in the hobby field. The instructions with starter sets are written for people who've never seen the equipment and have probably never built any sort of model before.

Remember: *Read all the instructions first! Then follow the instructions while you build.*

When building, don't rush. Don't panic. Don't goof it up. Don't try to cover up a goof. Do things right the first time, and they'll work right for you. I've helped thousands of model rocketeers get started for a quarter of a century now. (That sounds like a long time, even to me!) I know you're anxious to get your model finished and into the air for its first flight. But if you don't take the time to build your launch pad, electric launch controller, and model rocket correctly, according to instructions, you're likely to be disappointed when you finally push that launch button.

Current beginner's models are marvels of top-notch product engineering based upon careful studies of beginner's requirements and capabilities and upon thorough field tests of the models in the hands of beginners. Just because a model happens to be a simple beginner's bird, don't get the idea that it won't perform. Many beginner's models have contest-winning performance because of their simplicity. I have set national model rocket performance records with beginner's model kits.

You're likely to find ready-to-fly (RTF) all-plastic model rockets in hobby and toy stores. Some of these fly very well, but usually good flight performance has been made secondary to appearance for sales appeal, ease of manufacture, or sheer gimmickry. Such RTF model rockets are intended for people who just want something to occupy a Sunday afternoon, not for someone like you who's interested enough in model rocketry to pick up this book and read this far. Some RTF models are good for demonstrations. But you'll find it's much more fun and satisfying to build your own model rockets.

Although it often takes months to build a contest-winning scale model, most people can assemble a beginner's model with an all-plastic tail-fin unit in about an hour. The speed record for assembling such a model from the instant of opening the box to the moment of launch is held by Greg Scinto of Stamford, Connecticut, who achieved this dubious honor by doing it in 11 minutes 56 seconds back in 1970. He would have done it in less time but he accidentally glued his finger into the body tube while holding the shock cord mount in place for the glue to dry. (I've seen a lot of funny things happen, and from time to time in this book I'll bring one of them in to make a point or just to remind you that this is a serious book about having fun.)

Most model rockets, regardless of their details of construction, size, and performance capabilities, have the following parts and assemblies (See Figure 1-4):

1. A hollow plastic or solid balsa nose that fits onto the front of the hollow rocket body and that will come off.
2. A hollow, lightweight, thin-walled plastic or rolled paper body tube that is the main structural part of the model rocket airframe. This body tube holds the recovery device, the motor, and the fins.
3. A launching lug, which is a small tube like a soda straw that is attached to the exterior of the body tube. It slips over the launching rod of the launch pad, thus holding the model upright on the pad before launch and guiding the model during the launch phase of the flight.
4. A recovery device, such as a plastic parachute or paper streamer, that is packed inside the body tube and ejected forward from the body tube by a retro-thrust action of the model rocket motor at a predetermined time after launching.
5. The expendable, replaceable solid-propellant model rocket motor and its associated thrust mount and retainer.
6. Fins made of balsa, plastic, cardboard, or pressboard that are fastened on the rear end of the body tube and, like feathers on an arrow, keep the model traveling in a true and predictable flight path.
7. Recovery wadding to protect the recovery device from the gas of the retro-thrust action of the motor and to help form a piston to eject the device from the front of the body tube.
8. An electric igniter to start the solid-propellant model rocket motor.

The model rocket kit that you buy or that comes with your starter set will have all of the necessary parts for you to assemble a model rocket with all of these features. It will be up to *you* to assemble them into a strong, lightweight, streamlined model that will slip through the air at speeds up to 250 miles per hour (mph) without breaking apart because of air resistance. You will be assembling a model that flies faster than the fastest model airplane!

Don't let this shake you up. You can do it successfully *if you read and follow all instructions!* Millions of other people have done it. And today's beginner's models are a snap to build and fly. Just take your time, read the instructions, believe the instructions, and do it right the first time. Once your model rocket is in the air, you can't call it back to make a correction.

Many tyros don't bother to paint their first model rocket. Some beginner's models don't require painting at all. But paint it anyway for a very good reason: If you paint it a bright color such as vivid red, bright orange, or fluorescent orange, you will be able to see it better in the air during flight *and* in the tree in which it will inevitably land. You will also be able to find it more easily on the ground should it happen to elude the clutches of the rocket-eating trees that exist *everywhere* in the world. (Trees love model rockets and eat them very slowly, savoring them for as long as several months. You

Figure 2-2 Two of the best beginner's models are the Centuri Screaming Eagle (left) and the Estes Alpha-III (right). Both have plastic noses and integral plastic tail assemblies.

can watch parts of your favorite model disappear week by week as the tree slowly devours it. This is a proven fact. At least, it's a well-known hypothesis among experienced model rocketeers.)

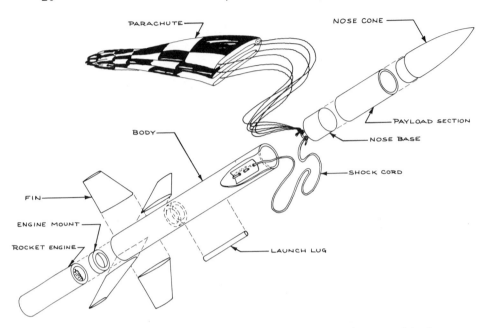

Figure 2-3 The basic parts of a typical model rocket are shown in this simplified disassembled view. The body tube is usually paper. The nose and fins
are balsa or plastic. Other parts are paper or plastic. The motor is always a
prefabricated, factory-made unit. This model has a small payload-carrying
section.

Nearly all model rockets operate in the same basic manner, although
performance can vary greatly due to differences in the weight, size, shape,
and power of the model rocket motor, and other factors that we'll discuss in
detail later.

When you're finally ready for that first flight session, excitement will
reign supreme! But before you leave the house to go to the flying field, make
certain you have *all* of the following items:

1. Launch pad.
2. Launch rod.
3. Electric launch controller.
4. The safety key for the launch controller. Put the safety key on the end
   of a bright strip of cloth or ribbon so that you don't lose it.
5. Your model rocket. (Some people have forgotten them.)
6. Three to six model rocket motors of the proper size and power for your
   model as recommended by the manufacturer.

TABLE 2
Altitude Range of Average Model Rockets

A small model rocket and a large model rocket were used to calculate the range of altitudes possible with a given motor type and a given lift-off weight. All altitudes are given in feet. Air resistance (drag) is taken into account assuming that a reasonably streamlined model rocket is flown. Note on field sizes: a regulation United States football field is 50 yards wide and 100 yards long. This is a field 150 × 300 feet in size.

Motor	Lift-off Weight (ounces)	Altitude Range (feet)	Recommended Field (feet)
Type A	1	235 to 560	150 × 300
	2	170 to 260	150 × 300
	3	100 to 120	150 × 300
Type B	1	400 to 1,040	300 × 300
	2	370 to 760	200 × 300
	3	280 to 425	150 × 300
Type C	2	650 to 1,620	400 × 400
	3	600 to 1,270	350 × 350
	4	520 to 800	200 × 300
	5	460 to 580	150 × 300
	6	330 to 420	150 × 300

7. An electric igniter for each model rocket motor plus two or three spares JIC (Just In Case).
8. A roll of paper tape such as masking tape.
9. A roll of cellophane tape.
10. Plenty of recovery wadding.
11. The ignition battery.
12. Your father or another adult.

Why this last item? Why your father or another adult? Simply to insure that you're successful. Flying a model rocket isn't difficult or complicated. But it requires that you do *everything* correctly in the proper sequence *before* you push the launch button. In that respect, it's just like a big rocket launch at the Cape. I highly recommend adult participation in the construction and flying of model rockets because I know too well from long experience that a young model rocketeer's natural enthusiasm and excitement over flying a model rocket can often cause him (or her) to overlook some important point in the countdown sequence. Although many young people are perfectly com-

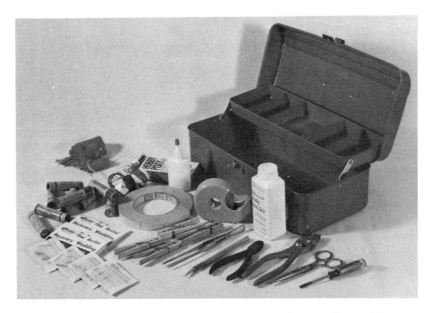

Figure 2-4 In addition to a model rocket, launch pad, controller, and battery, the fully prepared model rocketeer always carries a complete range kit with all the tools, pieces of equipment, and other goodies that make flying easier and more fun.

petent to build and fly a model rocket all by themselves in complete safety, and although they may know much more about it than an adult supervisor, the double-check feature of adult participation can often prevent mistakes made in haste and excitement.

The professionals use a double-check system at the Cape for the big rockets. Why should model rockets be any different in this respect from the big ones just because of their size? Besides, it's kind of fun to know more about something than one's parents.

Your next task is to choose a flying field. It should be away from major highways and freeways, power lines, and tall buildings. (How would *you* like to be buzzing along the freeway at 55 mph and suddenly see a rocket underneath a parachute descending right in front of your windshield?) The size of the flying field depends on how high you expect your model to fly. For most beginner's models propelled by Type A, Type B, or even Type C motors, a ground area about the size of a school athletic field is adequate on a calm day. On a windy day, the model will drift farther in the wind and will need some clear downwind area to land in . . . unless you want to feed rocket-eating trees.

To calculate the size of the field you should use, divide the expected altitude of the model by four. The launch field should have no ground dimension less than one-fourth the anticipated maximum altitude to be achieved.

Roughly speaking, a typical beginner's model with a Type A motor will go about 500 feet high; it will go about 900 feet high with a Type B motor and about 1,400 feet high with a Type C motor. This means that for flying with Type C motors you will need a field with the shortest dimension of about 350 feet, which is a little longer than a 100-yard football field unless there are sizable end zones.

Set up your launch pad on the *upwind* side of the field so the model will drift back across the field and land within its boundaries.

Important: After setting up your launch pad and electric ignition system, *test it.* Hook up one of the spare igniters that you brought. The igniter should *not* glow red-hot or activate until you have put the safety key into the launch controller and pushed the launch button. If your system can't pass this simple test, you have problems. You'll have to troubleshoot the electrical system to find out what's wrong. If the igniter doesn't glow red-hot or activate after you have inserted the safety key and pushed the launch button, it probably means that your battery is too weak or dead. This rarely happens today when using the sensitive new igniters that can be activated by several penlight batteries, but it's a common trouble point with older nichrome wire igniters. However, failure of the igniter to activate could also mean that you've got problems in the wiring. Read the chapter on ignition and launching for details about troubleshooting. If your electric ignition system successfully passes this pre-launch test, you can have confidence that the model will not take off while you're hooking it up—and that it *will* take off when you want it to!

Now *prep* or prepare your model for flight. Insert the recovery wadding. Fold up the recovery streamer or parachute so that it slides *easily* into the body tube. Stuff the shock cord in on top. Put on the nose, making sure that you don't jam the shock cord between the body tube and the nose base. The nose should be able to slide off easily.

For your first flight, use the lowest power motor recommended for your model. Usually, this will be a Type A motor. Check the instructions to make sure what kind of motor to use. Some larger beginner's models require a Type B motor for the best flight. Nevertheless, use the *lowest* possible motor power. The first flight is a test flight. You'll want to get the model back so you can fly it again. (Any fool can ram a Type C or larger motor into his model and lose it on the first flight. That's no outstanding achievement.)

Make certain that the motor cannot slide out of the rear end of the model. Most tyro models use a motor clip that holds the motor firmly in the body tube. If your model doesn't have a motor clip, wrap the motor with cellophane or masking tape until you have to force it into place. Be careful! Too much tape and too tight a fit may cause you to buckle the body tube when trying to get the motor in or out again later.

Install the igniter in the nozzle of the model rocket motor, following carefully the instructions that come with the motor package.

Now your model rocket is ready for its preflight safety inspection, given by an adult who's supervising the flight activities. Why bother? Simply because it is too late to correct a mistake after you have launched the model. It will be on it's way skyward, and you can't possibly catch it!

During all preflight times, keep the safety key to the launch controller in your pocket or hand. This insures that nobody but you will be able to switch the electricity into the motor igniter. You will be sure that the circuit is safe as you hook up the igniter.

Slide the model down the launch rod. The launching lug should slip easily over the rod. The function of the launching rod should be obvious to you by now. It supports the model during flight operations and hookup. And it provides the initial guidance of the model after ignition while the model's airspeed is still too low for the fins to stabilize the model.

Clear the launch area. Hook up the igniter, clipping one of the microclips on the firing leads to each end of the igniter. Make sure you have good connections. Make certain that the clips do not touch each other and that both of them are not shorting out through any metal launch-pad parts. Get in the habit of keeping your fingers out from under the direct line of the exhaust jet while doing this. Also, never look down on the model as it sits on the launch pad, hooked up and ready to fly.

Now retire to the launch controller about 15 feet away. Check the area. If it's clear of people and all other interference, the adult safety officer should give you the "all clear" or "all systems go" for launch. Insert the safety key. The ignition continuity light in the launch controller should come on, telling you that you have a hot circuit ready to launch. Give the old countdown:

"Five . . . four . . . three . . . two . . . one . . . start!"

When you press the launch button on the controller, the electric current will flow from the battery to the igniter in the rocket motor. The electrical resistance of the igniter will cause it to glow red-hot instantly (or less than instantly if you have a weak battery). This hot wire ignites the solid propellant in the model rocket motor.

The solid propellant begins to burn, creating large volumes of hot gas that rush out the rocket nozzle at more than twice the speed of sound, producing the thrust force that accelerates the model on its way.

The model will lift off very quickly. It will reach the end of the launching rod in about 2/10 second and be traveling about 30 mph as it takes to the air.

Powered flight lasts for 1 second or less. During this time, all the solid propellant in the motor is used up. At this "burnout" point in the flight, the model will be 50 to 200 feet in the air and traveling at a speed between 100 and 300 mph straight up.

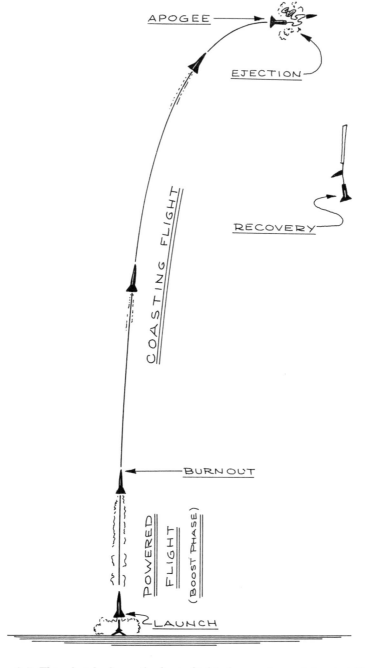

APOGEE

EJECTION

RECOVERY

COASTING FLIGHT

BURNOUT

POWERED FLIGHT (BOOST PHASE)

LAUNCH

Figure 2-5 This sketch shows the basic flight phases of a model rocket. This is the sort of performance you should expect of your first model rocket—and every one you build thereafter, too.

Figure 2-6 Lift off! The moment of truth arrives as you push the ignition button and your model rocket starts to streak aloft from its launch pad. (Estes Industries, Inc.)

Now coasting flight begins. The model trades speed for altitude. The end of burning of the solid propellant in the motor has started a slow-burning time delay charge in the motor. This time delay produces no thrust and allows the model to coast upward, delaying the ejection of the recovery device until maximum altitude has been attained.

At or near apogee—the maximum distance from the earth—and at a predetermined time after ignition chosen by the rocketeer who chooses the model rocket motor with the proper time delay built into it, the recovery ejection charge in the motor activates. This produces a retro-thrust puff of gas which pushes forward into the body tube and pressurizes the tube, forcing the wadding and recovery device forward and dislodging the nose. The recovery device deploys into the air and slows down the model. The model then returns slowly and gently to the ground. Since all parts are tied together by the shock cord, the entire model lands together where it can be easily recovered—unless it has (naturally) landed in the one and only tree in sight.

If the model lands in an electric power line, forget it. Don't ever try to get a model rocket out of a power line. You can buy a new model rocket for a couple of dollars. You cannot buy back your life.

With your model happily in hand, return to the launch area. Check the model for damage such as broken fins or cracked parts. Take out the expended model rocket motor casing and put it in your field box; take it home and throw it away later. Don't give it to the little urchins who will immediately come out of the bushes within minutes after your first launch; these kids may try to stuff the used casing full of match heads and get hurt. Repack the recovery device, install a new model rocket motor, and you're ready to fly the bird again.

I have seen model rockets that have made more than one hundred flights—but they didn't make a hundred flights in one afternoon. You'll be doing well to get off three or four flights in your first flight session. Try some higher powered motors and see the difference in performance of your model. Experiment with tilting the launch pad a few degrees away from the vertical.

Once you have put in your first afternoon of flying model rockets, you're ready to go on to bigger and better models. I'm certain that you'll be hooked on model rocketry at this point. Your dad may discover that he's hooked, too! I've known many fathers who grumpily went to the flying field for the first flights because their young rocketeers insisted on doing things the right way, and I've seen those fathers come home afterward far more excited about building and flying model rockets than their offspring. Model rocketry is not just a kid's hobby.

A model rocketeer's first flight is an important experience, and he or she remembers it even after flying thousands of models over the years. I can still remember my first flight on a chilly February morning in a cotton field near Las Cruces, New Mexico, in 1957. Today, many thousands of flights later, I get the same kick out of pushing that launch button and watching one of my creations lift off into the sky.

Sure, it won't get to the moon—but I can imagine it does!

3

Tools and Techniques in the Workshop

In working with model rockets, common sense is required. If you don't have it to start with, you develop it very quickly. You can develop your manual dexterity and your ability to work with tools as well.

The most important rule for becoming a successful model rocketeer is simple:

Read and follow all instructions completely.

The second important rule is:

If at first you don't succeed, try reading the instructions.

I cannot emphasize these two rules too strongly. The biggest mistakes made by novice model rocketeers (and even by some expert model rocketeers who think they know better but don't) is failing to read and follow instructions. As an indication of what can happen when a model rocketeer gets rushed or slapdash in his work or becomes a victim of "I-know-it-all," I have actually seen the following gross mistakes in beginner's model rockets:

1. A nose glued to the body tube so that the nose could never come off for the recovery device to deploy.
2. A motor mount glued in backwards so that it was impossible to insert a model rocket motor of any type into the model.
3. A solid bulkhead glued across the inside of the body tube so that no ejection charge gas could eject the recovery device.
4. Recovery wadding glued into the body tube.
5. Balsa fins that were cut out with a pair of scissors. "But, Mr. Stine, it makes such a nice, fuzzy edge for gluing!"
6. A model rocket motor firmly glued into the model.
7. An endless, disheartening parade of model rockets with crooked fins, fins that were too small, fins glued on the nose of the model (a definite no-no), no launch lug, launch lug mashed flat, no motor mount, shock cord not glued in, and (yes!) body tube crimped or bent.

Such models are garbage birds good only for tossing into the trash can. I know that *you* can do better!

Once you have built and flown your first beginner's model, you should go on to more difficult kits and, eventually, to designing your own models. First of all, give up every notion you may have had about rockets. Take that design for the four-staged radio-carrying mile-high supersonic rocket and tuck it away in your notebook; you aren't ready to build it yet. You have a lot to learn, and you'll have a lot of fun learning it. Later, when you look with greater knowledge and experience at that "early, primitive design," you will either laugh yourself silly or discover that it needs a lot of changes based upon what you've learned. This is true even if you're an experienced professional rocketeer working for NASA or the USAF who has taken up model rocketry as a hobby, or if you're already a championship airplane modeler who's decided to try something new.

We do many things differently in model rocketry, and we do them for a definite reason and for a specific purpose. Sometimes, we've learned about them the hard way. There's no need for you to repeat the past mistakes of others. In this book, you'll learn *why* we do things the way we do. You'll also be able to see why for yourself in your own workshop and out on the flying field.

WORKSHOP

First, you must have a place to work. Beginner's models and some other simple kits don't require much of a workshop. But if you want to progress beyond simple models, you'll have to develop a workshop. This is a place where you can keep your tools, your model rocketry supplies (spare noses, body tubes, balsa, and all the other things that you'll accumulate), your models as you're building them, and your models once they're finished.

A card table in your bedroom is all that you may need to start with. But balsa dust, wood chips, paint overspray, and other debris accumulates while you're building model rockets, and you really need a place that will keep this stuff from spreading itself thinly all over the rest of the house to the dismay of mother or wife. In addition, modeling paints and glues tend to be somewhat aromatic. In other words, to a person who isn't a modeler, they stink.

If you can manage to commandeer a corner of the basement as a workshop, be sure there are windows that you can open to provide ventilation. The days of glue-sniffing are over now because the glue manufacturers have included in the glue a substance that makes you violently sick to your stomach if you inhale too much glue vapor. The vapors of drying paints aren't good for you, either. So make sure you have adequate ventilation.

Garages make excellent workshops in warm climates where you won't freeze in the wintertime. If you live in a cold climate and can get a garage

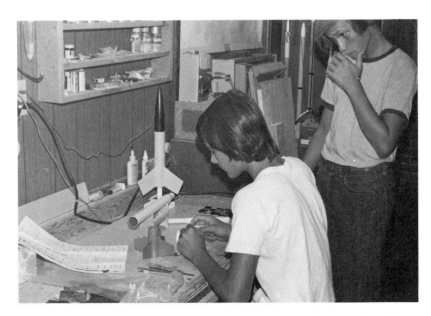

Figure 3-1 A well-lit, well-equipped workshop is essential for good modeling.

Figure 3-2 You can set up a simple workshop on a card table in your bedroom, but be sure to protect furniture and refrain from making lots of balsa dust.

as your workshop, it should be heated because it's difficult to do precision modeling work with cold fingers or while wearing gloves and other cold-weather clothing. But keep paints, glues, model rocket motors, and spray paint cans away from the heater unless you want to put the garage into orbit.

Some of us, myself included, are lucky enough to have a large workshop because the woman of the house is convinced that model rocketry is a great father–son activity that keeps the men happy and out from under foot. (It's been a great father–daughter activity for me as well because my two daughters were involved in the hobby when they were young; both became United States Junior Champions and one was a member of the United States team at the First World Championships in Yugoslavia and won a bronze medal there.) Some households boast having the whole family involved in model rocketry, which is an ideal arrangement for getting the best possible workshop area in the house.

Wherever you manage to set up your workshop, try to keep it in some semblance of order and cleanliness, impossible as that may be at times. This keeps you from tracking balsa dust into the house. It also permits you to find things more quickly when you're in that panic-rush to complete the model the night before a major contest. It also helps in finding that little bitty part which obeys Paul Harvey's Law: A dropped part will always roll into the most inaccessible part of the workshop. Mostly, however, a neat and clean workshop makes it easier for you to get along with the other nonmodeling members of your family.

SIMPLE TOOLS

Your most powerful tool is a notebook in which you can write down your ideas, make sketches, note how you did things so that you can remember them better in the future, and file technical reports and other papers. I have more than a dozen loose-leaf notebooks, each of them crammed with information garnered over the years. They are the most important books in my library. *The Handbook of Model Rocketry* was compiled from the information I'd tucked away in them. Scientists and engineers always keep notebooks and logbooks for their ideas, progress notes, reports, and other data. Model rocketry is a scientific and technical hobby, so it will pay you to start keeping a model rocketry notebook from the very start. There's a tremendous amount of information and data available in model rocketry, far more than in most hobbies. You won't want to lose it once you've got it, so put it away where you can find it again—in your notebook. Your notebook will keep you from "reinventing the wheel," which is duplicating somebody else's mistake that you should have read about, noted, and filed away to keep you from doing it, too.

Since you can't put a model rocket together with your bare hands, you'll need a few simple tools, and you'll have to learn how to use them. You may already have some model rocketry tools. All of them can be purchased quite inexpensively at your hobby shop or hardware store. (Some model rocketeers are tool collectors, too. The tool department of a hobby shop is akin to a candy store for these people.)

The use of tools is one of the many things that sets human beings above the beasts of the jungle. So get good tools to start with, treat them properly, and use them safely.

You'll need a *modeling knife.* This can be a new single-edged razor blade; don't use a double-edged razor blade because you can easily cut yourself with it. You can buy a modeling knife at a hobby shop for a dollar or so, and it will last for years. (I bought mine in 1941.) It has a slim metal handle with a holder, or chuck, at one end to grip a special, extra-sharp replaceable blade that will cut plastic, paper, and balsa. A typical modeling knife is called an X-Acto® knife. One of the best knives for modeling is the X-Acto No. 1 knife. I like to use it with the X-Acto No. 11 blade. Buy a package of extra blades for your knife so you'll always have a sharp blade. When the blade gets dull and won't cut neatly, remove it from the handle and install a new blade.

NOTE: When you use a modeling knife, be careful. It cuts fingers more easily than it cuts anything else.

A *pair of tweezers* is handy for holding parts that are too small for your fingers to hold; thus, they become a very fine extension of your fingertips. You can also use them to reach and grasp in small places. Tweezers can be purchased in a hobby store. You may also wish to get a pair of tweezers that *unclamp* as you squeeze them so that they can be used as long, thin clamps.

A sharp *pencil* or a *ball-point pen* is a must for making notes, marking parts, etc. To keep it from "walking away" from your workbench, keep it on a leash; tie or glue one end of a 24-inch length of string around it and attach one end to your workbench with a tack, nail, or staple.

Scissors are useful for cutting out paper templates, decals, and other paper parts. As mentioned before, please don't use them to cut balsa. You may also have to attach your scissors to your workbench with a string so that they don't get appropriated by someone else.

Needle-nosed pliers are useful for assembling things with nuts and bolts, for holding parts, and for bending metal. Again, your hobby store is a source for these.

A *small screwdriver* is a model rocketeer's friend in the shop and in the flying kit that you take to the flying field.

A metal 6-inch or 12-inch *rule* will be used to measure and to guide you in cutting straight lines with a modeling knife. Don't get a plastic or wood rule; you'll end up cutting its edge and will therefore make balsa cuts that are as crooked as a mountain highway. To cut a straight line with a

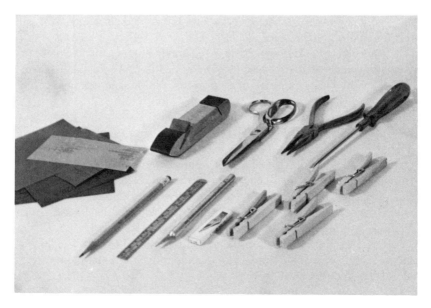

Figure 3-3 These are some of the simple, everyday tools that you'll use most in building all sorts of model rockets.

modeling knife and rule, merely lay the steel rule down on the balsa, cardboard, or paper, and run the knife along the rule edge.

If you can find a metal rule with both English system inches and metric system millimeters or centimeters on it, you'll discover how easy it is to measure things in the metric system. It's also easier to work in the metric system when building model rockets. A millimeter, for example, is 40/1000 of an inch (0.040 inch), or roughly 25/64 inch. It is much easier to work in the round numbers of the metric system than in the fractions of an inch of the English system. For example, it is easier to work with 15 millimeters than with its close equivalent, 19/32 inch. Model rocketry is technically on the international metric system and has been since 1962. However, many measurements are still given in the English system of inches, pounds, etc., because most Americans are familiar with these units. In model rocketry, you can become quickly and easily acquainted with the metric system.

A good stock of *sandpaper* in various grits should be kept at hand. It is used for shaping and smoothing. It comes in various grades ranging from very coarse to very fine. For model rocket work, you can buy large sheets of sandpaper at the hardware store at very low cost. Cut these big sheets into little sheets about 2 inches (50 millimeters) square, using a pair of scissors. Get sheets of No. 200, No. 320, and No. 400 wet-or-dry sandpaper. The

No. 200 is useful for fast shaping, the No. 320 is good all-around stuff, and the No. 400 is great for final smoothing.

You can buy a *sandpaper block* at the hobby store, or you can make one by taking a convenient-sized wood block and tacking some No. 200 sandpaper to it. The wood block provides a flat, firm base for the sandpaper and permits you to shape flats, curves, and weird shapes more easily.

You'll need some *clamps* to hold parts together while they're drying, to hold parts generally, and to use around your launch pad on the flying field. The finest modeling clamp known to man is the good old-fashioned spring clothespin. Before most people had clothes dryers, these clothespins were common household items. Today, you may have to buy some in the hardware store. Once you have them, you'll find all sorts of uses for them. Since they're made of wood, they can be sawed, drilled, carved, sanded, painted, glued, and made into many needed items.

These are the basic tools of the model rocketeer. I've built a lot of good model rockets in a one-room studio apartment or in a motel room with no tools other than these.

THE IDEAL WORKSHOP

Every model rocketeer dreams of the ideal workshop. I have one, but it took years, plus some cash, to accumulate all the tools, jigs, fixtures, and gadgets.

Probably the most ubiquitous power tool in American homes is the 1/4-inch or 3/8-inch *electric drill*. This is a very handy model rocket tool if you use some ingenuity to set it up in various ways. Sure, it will always drill holes if you put a drill bit in the chuck. But by using a drill press accessory available where you bought the drill, you can turn a simple electric drill into a very effective and accurate drill press, buffer, polisher, and grinder. With a horizontal holder, the electric drill becomes a grinder or a useful miniature lathe for turning model rocket parts out of balsa. Many modern electric drills have various speed controls and reverse features; they shouldn't be called "electric drills" because they are, in reality, portable rotary electric power sources. Think of your electric drill in those general terms, and you can find lots of uses for it.

If you have a small wood lathe or can get access to one in a school shop, for example, you're really in the model rocket building business! Such a lathe can be used to make many model rocket parts. Although the Austrian "Unimat" lathe is expensive, it is extremely accurate and will last for a long time. I bought one in 1957 and have used it in my model rocket work ever since; it has paid for itself many times over because I've been able to make a lot of special parts instead of buying them. There are other small, inexpensive wood

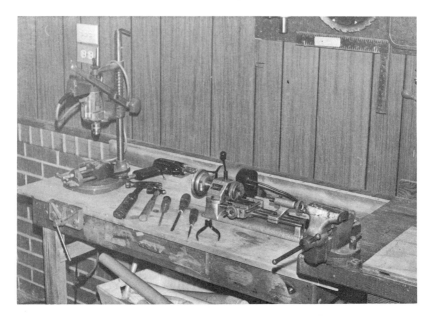

Figure 3-4 A complete workshop should include a drill press, small lathe, bench vise, and other handy tools.

lathes now available. Of all the power tools in a model rocket workshop, the lathe is perhaps the most used and most helpful.

Other handy tools include a *claw hammer*, a small *ball peen hammer*, *standard slot and Phillips-head screwdrivers* of various sizes, *side cutters*, *diagonal cutters*, *soldering iron*, *model maker's tee-head pins*, *a vise*, and *a building board*.

A building board is a rectangular piece of plywood, composition board, Homosote® board, cork, drywall, or the cutout from a Formica® countertop available at some lumberyards. The board provides a flat, smooth surface to work on, one that you can cut on without scarring the tabletop. You can easily stick pins into some of these surfaces to hold parts down against a drawing.

All sorts of little jigs and fixtures have been developed by model rocketeers over the years. The following are some of the ones I've found to be most useful.

The cradle stand—How do you set a model rocket down on the workbench while you're working on it? Answer: Do as the NASA sounding rocket people do and use a cradle. A typical cradle that can be cut from balsa sheet or from Bristol board or other stiff cardboard is shown in Figure 3-5. I started using cradles like this in 1957, copying them from the ones we were using

TABLE 3
Metric Conversion Chart

From	To	Multiply by
Centimeters	Inches	0.3937
Centimeters	Feet	0.0328
Inches	Centimeters	2.54
Inches	Millimeters	25.4
Inches	Meters	0.0254
Meters	Inches	39.370
Meters	Feet	3.2808
Meters	Yards	1.0936
Feet	Centimeters	30.48
Feet	Meters	0.3048
Feet	Kilometers	0.0003048
Grams	Ounces	0.03527
Grams	Pounds	0.0022046
Kilograms	Ounces	35.2739
Kilograms	Pounds	2.2046
Ounces	Pounds	0.0625
Ounces	Grams	28.349
Ounces	Kilograms	0.0283949
Pounds	Grams	435.592
Pounds	Kilograms	0.45359
Newtons	Pounds-force	4.45
Pounds-force	Newtons	0.2247
Pounds–weight	Pounds–mass	0.03105
Pounds–mass	Pounds–force	32.27
Miles/hour	Feet/second	1.467
Feet/second	Miles/hour	0.6818
Meters/second	Feet/second	3.281
Feet/second	Meters/second	0.3048

for the full-sized Aerobee sounding rockets at White Sands, New Mexico, where I was then working. If you have a model that is shorter or longer than the cradle I've shown in Figure 3-5, you can make a shorter or longer cradle. Or you can make it bigger for bigger models. I have a series of cradles ranging from 3 inches to 12 inches long. Sometimes, I paint them if they're to be used to display a model. You'll find that a cradle is one of the handiest fixtures in the shop.

The spike—Wendell H. Stickney built the first spike in 1961 to hold a model rocket vertically during construction or for display. The easiest way to

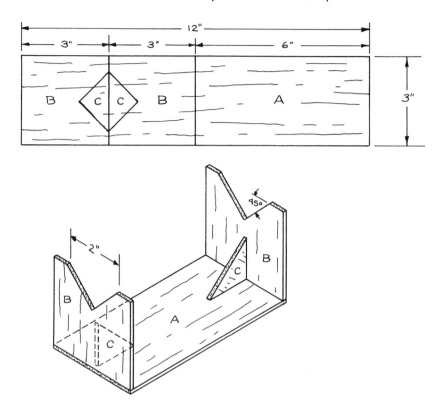

Figure 3-5 A cradle stand to hold model rockets in a horizontal position while working on them can be cut from sheet balsa or stiff cardboard as shown.

make a spike is to glue an expended motor casing to a block of wood. However, you can make a better one with a lathe to help out. A typical spike is shown in Figure 3-6. It's longer than a regular motor casing and will hold a model high enough to clear any swept-back fins that extend beyond the rear end of the body tube. A group of spikes will increase your workbench space because they hold models vertically.

Spike row—The obvious and logical extension of the spike is several spikes fastened to a long piece of board. This is useful not only in the shop or for an impressive display, but is handy when you have to carry a lot of models. It's as easy to make as a single spike. Most of my spikes and spike rows are now made with 1/2-inch wood dowels as the spikes; they'll fit inside models designed for minimotors. To use them with standard motors, slip an expended 19-millimeter standard motor casing over the spike, or use the motor mount location and spacer tube that comes with most kits. For use

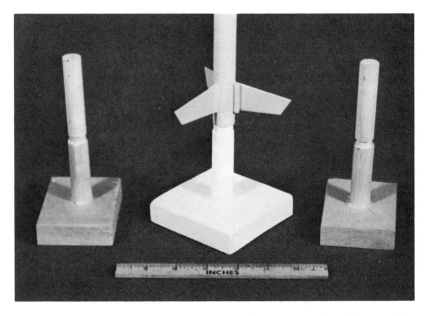

Figure 3-6 The workshop spike can be made from wood or plastic and is designed to hold the model vertically.

Figure 3-7 A spike row is a line of spike dowels on a single board for holding several models vertically.

Figure 3-8 A paper painting wand is made from a sheet of newspaper rolled up and stuck into the rear end of the body tube. It keeps you from painting your hand when spray painting a model rocket.

with models intended for motors with larger diameters, either make a bigger spike or take a standard spike and wrap paper tape around it until its diameter is built up to handle the bigger model.

The paper wand—How do you paint a model rocket without getting paint overspray all over the hand that holds the model? Make a paper wand by rolling up a sheet of newspaper and fitting one end of it into the rear end of the model. Hold onto the other end. Spray only at the model end, of course. The flimsy newspaper becomes quite strong after you've rolled it into a wand. Figure 3-8 shows a model being spray painted using a paper wand.

The rotating spike—For painting models with an airbrush, spray gun, or spray can, I built a spike mounted horizontally on a support that would permit the spike to be rotated horizontally. The spike is slipped into the model's motor mount. The model is thus held horizontally and can be rotated like a roasting pig on a spit over a fire. This permits easy spray painting of the entire model. Making the gadget requires some work with a lathe and drill press, but is not so complicated that you can't finish it in a couple of hours as a quickie project.

Fin alignment angle—How do you draw a line accurately along a model rocket body tube so that you can correctly locate the fins? The answer is so

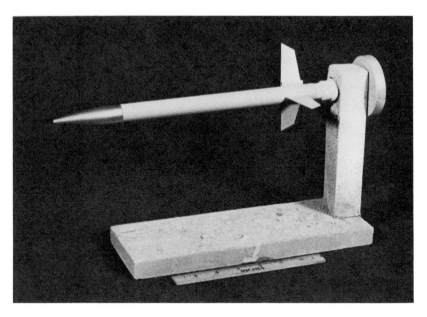

Figure 3-9 A rotating spike holds a model horizontally so it can be turned to spray paint all sides of it.

Figure 3-10 A small metal angle laid against the side of a body tube permits a straight line to be drawn down the tube for positioning fins.

simple you may not believe it. Go to a hobby or hardware store and buy a small piece of metal angle about 6 inches long. You should get 1/2-inch angle or smaller. If you can find some brass angle used to make slot car bodies, it's perfect. Simply lay the angle down along the body tube as shown in Figure 3-10. Use one edge of the angle as a ruler to draw a straight line along the body tube. The line will be perfectly aligned with the long axis of the body tube.

Fin positioner—In the history of model rocketry, thousands of different fin assembly jigs have been designed to help the beginner get the fins glued on straight and true. The best one I know of was invented by Howard R. Kuhn, many times United States Senior National Champion, member of the United States Space Modeling Team, and currently Chairman of the Space Modeling Subcommittee of the Federation Aeronautique Internationale. You can buy one of Kuhn's fin jigs from Competition Model Rockets at the address listed in Appendix 1. Or you can make one. Go to your hardware store or lumberyard; get a piece of wood angle as shown in Figure 3-12. Cut off

Figure 3-11 With a fin positioning drawing such as this, fins can be properly located on the end of a body tube.

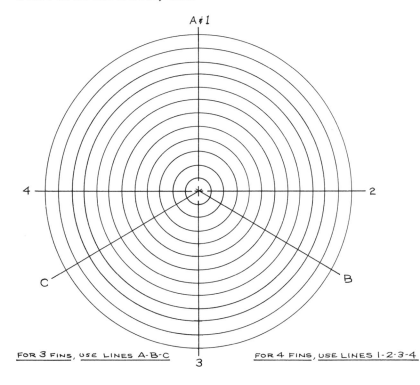

A & 1

4

2

C

B

FOR 3 FINS, USE LINES A-B-C   FOR 4 FINS, USE LINES 1-2-3-4

3

about 6 inches. With a saw, cut a slot at the apex of the angle as indicated. Make the slot 1/16-inch wide for normal fins; you can make another jig with a 3/32-inch slot for heavier fins, and so forth. Put the angle on the body tube as shown and hold it in place with two rubber bands around the body tube. Insert the fin through the slot, and the jig will hold the fin in perfect position until the glue dries. The design of the Kuhn jig eliminates the worry about gluing the jig to the model in the process.

Q–tips®—Often you'll need to put a bead of glue around the inside of the body tube in order to glue a motor mount in place. The easiest way to do this is with a tool swiped from the bathroom medicine cabinet and widely known by the trade name Q-tip®. It's a wood or paper stick with a ball of cotton on one or both ends. You can easily make one using a small wood dowel and cotton. Or you can buy a box of them at the drug store. Mark on the stick the distance inside the body tube that you want the glue applied.

Figure 3-12 The Kuhn fin jig can be made from a piece of wood angle and will hold a fin straight until the glue dries.

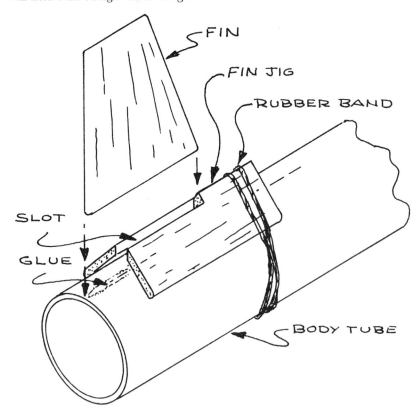

Figure 3-13 A cotton swab or Q-Tip® can be used to apply glue to the inside of a body tube. Mark the swab stick to indicate how far into the tube to apply the glue.

Put glue on the cotton swab and carefully insert the Q-tip down into the body tube, being careful not to get glue on the inside walls until the Q-tip has reached the proper insertion depth. Then swab the glue around the inside of the tube. It may take several applications before you get a good band of glue in there, but you can use the same swab many times. In fact, the swab gets better and better as you continue to use it because a hard glob of dried glue builds up on the cotton, making it easier to use.

There are many other special tools, jigs, and fixtures that have been developed by model rocketeers. I'll probably hear very shortly about all of those I haven't mentioned here. But I've told you about those items that I've tested and found to be most suitable and useful in the model rocket workshop.

GLUES

There used to be only one kind of glue used in building model rockets: model airplane glue. But the field of "bonding technology" has so progressed that many kinds of adhesives are available today. We call them "glues" although in the true technical sense many are "bonding agents." They give the model rocketeer a wide choice to use with many different materials.

The materials used in model rocketry are not difficult to join by bonding or gluing. But you must use the proper bonding agent. Porous materials such as paper, cardboard, and balsa require a different kind of bonding agent than nonporous materials such as plastic and metal.

The following information on glues and bonding agents has been accumulated by years of model building and by correspondence with aerospace modelers from all over the world.

Model airplane glues, such as Testor's Formula B®, Pactra C-77®, Ambroid®, and Du Pont Duco Cement®, are used for gluing balsa to itself or to fiber or paper body tubes. They will also glue paper together. They will bond some plastics but are not considered optimum for use on plastics because they will often melt some kinds of plastic.

White glues such as Elmer's Glue-All® are casein-based, made from milk, and are water-soluble before they set-up and harden. They are exceptionally strong for use with paper, wood, and other porous materials, but they will not work on plastics. Their major disadvantage is the fact that they take a long time to set-up firmly and even longer to dry completely—overnight in some cases. You may also have trouble applying some kinds of paint over white glue layers.

Aliphatic resin glues such as Franklin Titebond® and Centuri Superbond® look like brownish-yellow versions of white glue. But they have a different chemistry. They are very strong and will set-up and dry faster than white glues. They are nearly transparent when dry, and most kinds of paint will go over them nicely. Aliphatic resin glues are very useful on porous materials but will not bond plastics.

Contact cements include rubber cement, Goodyear Pliobond®, and Weldwood®. To use them, you have to be *quick* and *accurate*. Apply a coating to both surfaces to be joined and allow the coats to dry for a few minutes. Then *carefully* join the surfaces. Contact cements bond firmly *immediately* upon contact, so be careful because you don't have any leeway in moving things around to line them up. Contact cements will usually bond anything to anything else, but it's always best to run a test first using a couple of pieces of scrap material.

Epoxy bonding agents are now commonly used in model rocketry. They are strong, fast-curing, and universal in their bonding capabilities. They'll bond almost anything to anything else—except Teflon® and some kinds of polyvinylchloride (PVC). Basically, epoxy is a plastic technically known as *thermosetting*. Epoxies come in two parts: a resin and a hardener. When you're ready to use the epoxy cement, mix together an equal amount of resin and hardener (or mix according to instructions; some epoxies require measurement of the amounts) in the amount you'll need within the next few

Figure 3-14 A wide variety of glues can be used in constructing model rockets, but some are intended only for special applications.

minutes. There are some epoxy types that will cure in 60 seconds; they do not have the strength of epoxies that harden or "cure" in five minutes, and the strongest epoxies are those that cure in 1 hour. Mix only what you need for a particular joint because, once mixed, epoxy has a limited "working life" ranging from 30 seconds for the fast stuff up to 30 minutes for the long-term curing types. Epoxy bonding agents are incredibly strong. They will bond nearly anything to anything else and bond it strongly. The thermal protection tiles on the NASA Space Shuttle are bonded into place using epoxy bonding agents.

Cyanoacrylate bonding agents have revolutionized model building in the last decade. For a long time, there was only one of these available: Eastman "910"® adhesive. Now cyanoacrylate brands abound. Generically, they're called "hot stuff" although this is the trade name for one of the most popular model-building brands. This bonding agent is a liquid that bonds anything to anything. It cures or sets-up *immediately* by the application of *pressure* between the two parts to be joined by a film of cyanoacrylate. Pressure must be maintained for one to two minutes for the bond to set up. Full curing takes place within 24 hours. You can apply model rocket fins and other *external* parts very rapidly with this agent. And it's great for quick repairs on the flying field. WARNING: Cyanoacrylates will cause *immediate* bonding of human skin to itself or to anything else! Don't get it on your fingers, in your eyes, in your mouth, or anywhere else on your body! If you do bond your fingers together or to something else, first try soaking the bonded areas with Pactra AeroGloss Thinner® or nail polish remover; if this doesn't release the bond, medical surgery will be required to separate you from yourself or your model. Great stuff, but be careful with it.

Cements or glues for plastics work differently than glues for porous materials. Plastic cements actually soften and melt the plastic material so that when two plastic parts are pressed together, the plastic flows together and welds.

Professional plastic model makers use acetone or methyl–ethyle–ketone (MEK) which they buy in large cans holding a gallon or more. You can go this route if you have 15,674 plastic models to assemble; otherwise, buy MEK and liquid plastic cements in small bottles from the hobby store and keep the caps on tight when you're not using them. They're all highly volatile and will evaporate quickly if left open to the air.

Plastic cements also come in tubes like some kinds of wood cements. This sort of plastic cement is actually plastic material with lots of solvent in it. When the solvent evaporates, the cement becomes solid plastic. When you apply it to a plastic piece that will be joined to another plastic piece, the glue forms a bridge of plastic that unites the two parts. Most hobby plastic cements are intended to bond polystyrene, which is the plastic used in most plastic model kits. It doesn't work well on other plastics such as PVC. One of the very best plastic cements is Du Pont No. 9011 Plastic Cement®, which is the best type of plastic cement to use for gluing clear acrylic plastic fins to paper body tubes. This kind of plastic cement also works well to bond polystyrene plastic to paper or wood, but it takes a long time to dry and harden.

If you don't know whether the glue you have will work with certain materials, make some tests using scraps of the materials. Glue them together and see what happens. This doesn't take very long to do, and it may save you all sorts of grief—such as ruining your model during construction or having it come apart in flight at 300 miles per hour.

There is a humorous side to bonding techniques. In aviation, there is the speed of sound, and in physics, the speed of light. In model rocketry, there is a similar "barrier" speed known as the "speed of balsa." This is the speed at which balsa construction fails in flight. It is a "variable constant" because it depends totally on the individual model builder. Some people continually build model rockets that exhibit a very low value of the speed of balsa.

You can increase the speed of balsa tremendously by making a proper glue joint. Having a good glue is only half the battle; knowing how to use it properly is the other half. Almost 90 percent of the people who make model rockets don't know how to make a good glue joint, even though the instructions tell them how to do it. This is probably because they think the correct method is only a ruse to get them to buy and use more glue. Not so.

If you want to make a good glue joint that is stronger than paper or balsa, follow these simple instructions for *all* kinds of bonding agents:

When gluing together porous materials such as paper or wood, use a *double-glue joint*. Coat both surfaces with a layer of glue or adhesive and let it dry. Then coat both surfaces again and join them together. The first coat of glue on both surfaces penetrates the pore of the material. The second coat is then free to join with the first coat and with the second coat on the other surface. A double-glue joint will be so strong that the materials will break or tear before the glue joint turns loose. Try it. It will surprise you. And you'll never use any other kind of glue joint again.

When gluing two pieces of plastic or other nonporous material, use a variation of the double-glue joint. Coat both surfaces with a bonding agent. Since the surfaces don't have pores, there's no need to apply two separate coats of glue to each surface. But the cement will soften each surface or adhere to each surface so that the bonding agent will bond to itself when you bring the two pieces together.

Having now become an expert in workshops, tools, and construction materials, you can tackle the job of building model rockets with considerable confidence that *perhaps* you can beat Murphy's Law—"If anything can go wrong, it will"—if you practice what you've learned.

4

Model Rocket Construction

Today, most model rockets are built from kits that contain all of the parts and components necessary to complete the model. Many of these parts are prefabricated, and a model rocketeer really isn't required to complete any highly detailed or difficult parts that call for special techniques or power tools such as a lathe. However, there is a growing cadre of dedicated—some would call them "crazy"—model rocketeers who prefer to build their model rocket airframes from scratch. Sometimes, they'll use parts available from the manufacturers, but otherwise they prefer to make everything themselves except for the model rocket motors; making model rocket motors in your basement or garage isn't purist; it's suicidal.

However, it doesn't make any difference whether you're a maker of prefab kits or a "scratch builder." Certain universal techniques are used by both kinds of space modelers.

I'm devoting an entire chapter to this general subject because, in my large collection of hobby how-to books, most authors assume that their readers are experienced and don't need basic explanations of how and why. I find the *why* to be very important to the technically curious person who becomes involved in model rocketry.

So, back to Square One.

NOSES

The nose is the front end of a model rocket airframe. After this astounding and illuminating statement, let me add that its shape can vary widely. We used to call it a "nose cone," but since its shape was rarely that of a true cone, we soon started calling it just a "nose."

Some of the most common kinds of nose shapes are shown in Figure 4-2.

Figure 4-1 Assembling your own model rocket either from a kit or from basic parts to your own design requires some craftsmanship that will improve as you build more models.

Figure 4-2 Some common model rocket nose shapes.

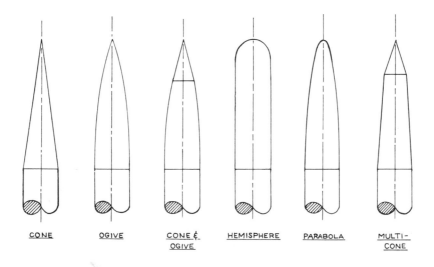

Unless your model rocket uses a special kind of recovery device such as rearward ejection (See Chapter 12, Recovery Devices), the nose must always be free to slide forward and come off. Therefore, the back end of the nose is cut down to form an internal shoulder or, technically, a tenon, that will slide inside the body tube and hold the nose in place.

The base diameter of the nose should match the outside diameter of the body tube. The diameter of the shoulder or tenon should be slightly less than the inside diameter of the body tube (about 0.005-inch to 0.010-inch smaller) so that it has a slip fit inside the tube. Remember, it's better to have the nose tenon a little loose because you can always build it up or "shim" it with cellophane or paper tape wrapped around the tenon to exactly the correct fit. If you have a nose tenon that's too big for the body tube, and if you've got a balsa nose, it's possible to roll it down, or "scrunch" it, by rolling the tenon along the edge of a table, crushing the balsa down until you get a proper fit.

There are a very large number of different noses now available from model rocket manufacturers.

Noses are usually made from turned balsa wood or from injection-molded or blow-molded polystyrene plastic. *Never* use a metal nose! If you must increase the weight of a nose to achieve the proper stability for your model, as we'll discuss later, use a flat metal nose weight, plasticene clay, or drill a hole in the base of a wood nose and fill it up with clay or glazier's lead putty.

Figure 4-3 Some of the body tube sizes available from Estes Industries, Inc. Other companies have a large selection of body tubes available, too.

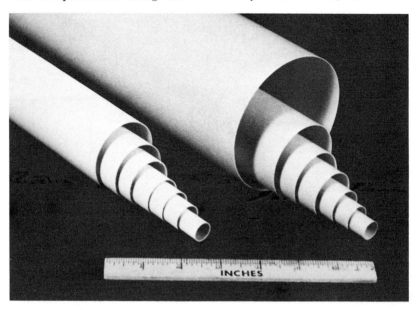

Don't stick a metal pin or nail into the tip of the nose to simulate a nose probe antenna. It could become a rocket-powered dart if something goes wrong in flight. You don't really have to have a sharply pointed nose on your model rocket; slightly rounded noses work better, as we'll see later.

You can make your own special noses if you have an electric drill in your workshop. Drill a 1/4-inch hole into one end of a balsa block. Glue a 1/4-inch hardwood dowel into the block so that it protrudes about one inch. This gives you something to tighten into the chuck of the electric drill (balsa is far too soft to hold well in a chuck). When the glue dries so that the balsa block doesn't separate from the dowel and chase you around the workshop, chuck the dowel in the drill chuck, turn on the drill, and proceed to *carefully* carve the balsa down to the desired nose shape using very coarse sandpaper. This is a tough job to do because the block starts off square, and you've got to turn it into a cylinder first. Be prepared to create some unusual egg-shaped noses at first.

The ultimate is, of course, to turn the nose on a precision lathe. You may need to do this when you get into scale model work and find yourself required to hold dimensions as small as a thousandth of an inch with balsa. It can be done. It has been done. But it isn't easy, and beginners can rarely do it without a lot of practice beforehand.

BODY TUBES

Body tubes for model rockets are usually made from thin-walled paper tubes. Because such tubes are hard to make and even more difficult to find in stores, most model rocketeers buy them ready-made from hobby shops or by direct mail from the manufacturers.

Body tubes are available in diameters ranging from 0.197 inch to 6 inches or more. Common lengths are 18 inches and 24 inches. Buy body tubes in the longest possible length and cut them to the custom length you require. You may have some scraps left over, but you'll find plenty of uses for these as motor mounts, stage couplers, spacers, payload supports, etc.

To cut a tube to the required length, first measure the desired length on the tube with your steel rule and mark the tube. Then wrap a piece of paper or file card stock around the tube at the pencil mark and draw a line around the tube's circumference using the card edge as a guide. Cut the tube with a No. 11 X-Acto® knife blade in a holder. Take several passes around the tube, cutting only a little on each pass. Don't try to cut all the way through the tube on the first pass around, or you'll mess up the tube. After some practice, you should be able to cut a body tube so you'll never be able to tell that it was cut to custom length.

Bodies for scale models that are not cylindrical in shape can be made by the "hollow log" technique. A balsa body is first turned to external shape on a lathe. Once the external shape has been turned, the modeler has two choices of style in completing the body. Sometimes a hole can be drilled completely down the center of the body block so that a regular paper body tube can be inserted into the block; this tube then holds the motor mount, shock cord attachment, and recovery device like an ordinary cylindrical model. However, this technique usually results in a heavy model. The true hollow log technique requires you to cut the body block in two, lengthwise, with a very thin saw blade. Hollow out both halves until the sidewalls are about 1/16-inch thick, just enough so you can see light through the balsa. Glue a paper body tube down the middle of one of the halves. Glue the two halves back together again. With a little sanding and filling, you won't be able to tell where the joint is, and you'll have a very lightweight, thin-walled balsa shell for a body.

MOTOR MOUNTS

A motor mount in a model rocket serves two purposes: (1) to hold the model rocket motor firmly in place so that it cannot move forward under thrust or backward (and out of the model) upon activation of the ejection charge; and (2) to hold the model rocket motor straight and in alignment with the model rocket airframe so that the thrust is directed along the center line of the model.

If the motor is not held firmly in the model, the thrust can ram it forward so that it comes out the front end of the body tube. If this happens, the motor usually reams the body tube, taking the wadding, recovery device, shock cord, and shock cord mount along with it on a short upward flight. This isn't good for your model.

On the other hand, if the motor is not mounted firmly in the model, the ejection charge can pop the motor backward out of the model. When this happens, the recovery device usually doesn't deploy at all. The model comes down like a streamlined anvil, usually saving the modeler the trouble of disposing of it because it buries itself. These "death dives" are not funny, especially since they always seem to happen during a critical demonstration when you want everything to work perfectly. So use a proper motor mount.

A motor mount usually consists of several basic parts: (1) the motor-mount tube into which the motor slides with a slip fit; (2) the thrust-mount ring to prevent the motor from ramming forward during thrust; (3) centering rings that will center the motor mount in the larger body tube; and (4) usually a thin springlike motor retaining clip.

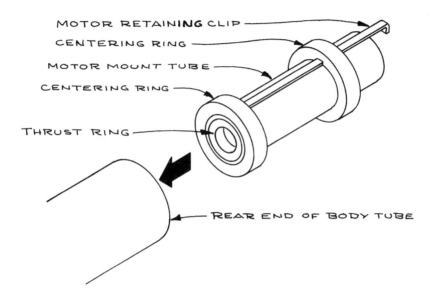

Figure 4-4 Drawing of a typical motor mount for positioning and retaining a model rocket motor in a larger body tube. Centering rings and motor tube can be eliminated if the body tube is the same size as the motor tube.

In simple models the body tube itself has the proper diameter for the motor to slip into without the need for a motor-mount tube and centering rings.

A thrust-mount ring is usually a small paper doughnut glued to the motor-mount tube to prevent the motor from going forward. It has a hole through its center to allow the ejection charge gas to pass. Often the motor retaining clip incorporates a small forward tab that serves as a thrust mount.

Centering rings are sometimes larger paper doughnuts or, for large body tubes, cardboard or tagboard disks with holes in their centers. Sometimes they are slotted to clear the motor clip. Although the shock cord can be attached to a centering ring, this is not considered good practice because it makes the shock cord susceptible to being burned through by repeated ejection charges.

Some models and motor mounts do not use motor clips, although a clip is almost 100 percent insurance that the motor will not be ejected in flight. A clip also makes the prepping of the model very fast and simple. If the model does not have a motor clip, you'll have to make sure the motor is installed tightly in the motor-mount tube. It should be tight enough that you cannot pull it out of the model with your fingers. But you should be able to remove it by giving it a firm, sustained pull with pliers. For this reason, you should always assemble the motor mount so that about 1/4-inch of the motor casing will protrude from the model so you can grasp and extract the motor with the slip-joint pliers you keep in your field box for just this purpose. If the motor casing is flush with the aft end of the motor-mount tube, you'll need to *push* the motor out with a pusher rod, a 12-inch to 18-inch length of 1/4-inch hardwood dowel that you also keep in your field box—or don't you? Just stick the pusher rod down through the body tube, being very careful that you don't push the *entire* motor mount out of the body tube in the process of pushing out the motor casing. Some motors become so fond of a model that they don't want to leave at all.

If you do a good job of assembling the motor mount and gluing it securely into the model, it will never come out. This is an achievement worth striving for.

FINS

A model rocket must have fins, or stabilizing surfaces, on its rear end in order to fly properly. You should not experiment to determine the truth of this statement. I'll prove it later.

The fins on a model rocket are like feathers on an arrow. They keep the model going straight in the air. Model rockets do not fly in the ordinary sense of the word. Their fins are not wings to provide forces to keep them aloft, but are stabilizing devices to ensure a straight and predictable flight path.

Some of the beginner's model rocket kits have molded polystyrene plastic tail-fin units that fit right over the rear end of the body tube. This eliminates the most difficult and time-consuming task of constructing a model rocket: getting the fins on straight and strong. If the fins are not attached correctly, the model can fly erratically or not at all. These plastic fin units provide the sort of true and predictable flights required of beginner's models.

With some plastic tail assemblies, it's vitally important to ensure that they don't slip off the rear end of the body tube when the thrust of the motor accelerates the model off the launch pad and into the air. When this happens—and it does to some beginners—you never want it to happen again. The easiest way to prevent it is to wrap one or more layers of tape around the rear end of the body tube, enlarging the tube just enough for the tail assembly to have a nice, snug fit over it. Many modelers glue the plastic tail assembly to the model; this insures it will not leave the party. But if a fin breaks on landing, the entire model must be thrown away because there is

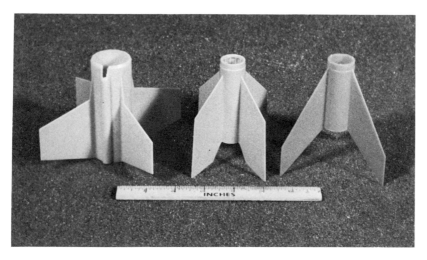

Figure 4-5 Some available injection-molded plastic tail assemblies with fins molded as part of the unit.

no way to replace the tail assembly. The best you can do is try to glue the broken fin back in place, and this is sometimes difficult to do.

Most model rockets, however, are built by cutting fins out of cardboard (for easy models) or sheet balsa. The fins are then glued on the body tube in the proper locations.

Some kit models come with die-cut balsa fins that don't have to be cut out with a knife. But it's always a good idea to run around the die-cut lines with a sharp No. 11 X-Acto® to ensure that the die has cut all the way through the balsa sheet and that the fins will pop out of the sheet without breaking in the process.

If the fin isn't die-cut, you'll have to cut it out of the sheet yourself. Model kits either have the fin planform or outline printed on the sheet balsa or on a paper pattern. In the latter case, you must cut the pattern out of the paper, lay it on the balsa, trace around it with a sharp pencil, then cut the fin from the balsa sheet with a knife.

Important: When laying-out a fin pattern on a balsa sheet, the grain of the balsa must always run parallel to the leading edge of the fin (see Figure 4-6) or *outward* from the body tube. If the balsa grain runs parallel to the body tube, the fin will not be strong enough and will be easy to break. Die-cut fins or printed-on-balsa fin patterns are already oriented with the balsa grain parallel to the leading edge of the fin.

To make sure you never forget this important point, here's a memory teaser to help you remember: Never forget that *the grain runs out!*

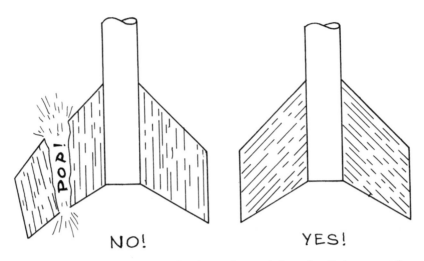

Figure 4-6 Unless you want balsa fins to be weak "pop fins," always cut fins so the balsa grain follows the leading edge of the fin pattern.

Cut the fins from the sheet balsa with your modeling knife. Try to cut squarely across the balsa sheet with the knife blade vertical. To cut a straight line, cut along the edge of your metal rule. Making a square, straight cut may be difficult to do at first, but keep at it. Each of us develops his own personal style of holding the knife and making the cut.

Large fins can be made stronger by using thicker sheet balsa. For most small sport models, 1/16-inch sheet balsa is usually strong enough. For larger models or for large fins, use 3/32-inch or even 1/8-inch sheet balsa. Sheet balsa in varying thicknesses from 1/32-inch to 1/2-inch is available in most hobby shops.

To further strengthen a large fin, glue paper on both sides of it. You can also cover fins in the time-honored model airplane method using tissue, silkspan, Monokote®, Coverite®, Solar Film®, or other kinds of model airplane coverings. This makes a sheet balsa fin very strong indeed. It also eliminates the balsa wood grain and makes the fin easier to finish smoothly.

I've built very strong fins using polystyrene foam plastic from packing crates as a core to give the fin its shape, then covering the foam core with Monokote® and shrinking it to make a modeler's version of the professional aeronautical engineer's stressed skin construction. This produced a very light fin with a beautifully slick finish and excellent strength.

Once you've cut out all the fins, stack them together and sand the stack to make sure all the fins have the same size and shape. This may not seem to be overly important with sport models, but it can become critical with

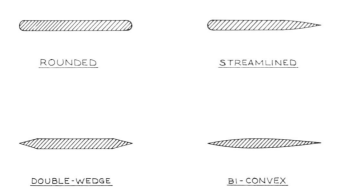

ROUNDED STREAMLINED

DOUBLE-WEDGE BI-CONVEX

Figure 4-7 Some basic model rocket fin airfoils shown in cross section.

high-performance competition models or with scale models. So develop this stack sanding habit from the very start in your model rocket activities.

Your model will fly okay if you simply glue on the fins with square edges. But performance and altitude can be nearly *doubled* if you take the time to put a streamlined airfoil on the fins. Some common ones are shown in Figure 4-7. The simplest one merely has the edges rounded—except at the fin root where you glue the fin to the body tube. You can round fin edges quickly and easily with your sanding block.

In the early days of model rocketry, many modelers put sharp-nosed wedge-shaped airfoils or sharp double-convex airfoils on the fins of their models, emulating the appearance of the bigger rockets at the Cape. But the "big ones" are designed to fly at supersonic and hypersonic speeds, while our model rockets rarely attain half the speed of sound and are subsonic vehicles. Therefore, the best airfoil for model rocket fins is one with a rounded leading edge and a tapered trailing edge as shown. It takes a little time to do this shaping with your sanding block. You may goof a couple of fins in the process, but cut some more and try again. The effort will pay off handsomely in improved performance.

Don't round or taper the fin edge that will be glued to the body tube! Like everything else, I've seen model rocketeers do it.

How many fins should you put on a model rocket? The bare minimum is three fins in the triform arrangement shown in Figure 4-8. Some models have the four-finned, or cruciform, arrangement. Models are also made with five and six fins, but rarely with more than six.

. Always put the fins at or near the rear end of the body tube. The reasons for this and the reasons why fins have certain sizes and shapes will be discussed in detail later.

You should *never* put fins near the nose of the model. Nor should you put any fins anywhere except near the rear end unless the kit instructions tell you to do so and tell you precisely where to put them. Things begin to get very tricky when you put the fins up front.

You can locate the fins on the body tube by using the fin placement guide as shown in Figure 4-9. Locate them carefully so they stick straight out from the body tube and are perfectly aligned with the model. If the fins are canted or crooked or cocked, the model may take off at an odd angle, spin, fly erratically, or otehwise act up in flight. If you put the fins on straight and true, the model will fly straight and true. Don't forget that the fins are the model's stabilizing system.

To get a powerful glue joint on a body tube, lightly sand off the top layer of paper on the tube where you're going to glue the fins. This removes the smooth, nonporous glazed outer layer of the tube that assists painting but

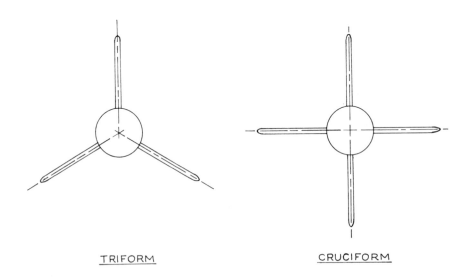

TRIFORM   CRUCIFORM

Figure 4-8 Rear view of two model rockets showing three- and four-fin configurations.

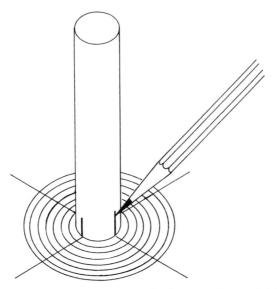

Figure 4-9 Using the fin placement guide shown in Figure 3-11.

hampers gluing action. It allows the glue to seep into the pores of the paper tube so that it can "grab hold and hang on."

Use a double-glue joint for attaching fins to the body tube. The fins will break and the body tube will tear before the double-glue joint cracks and turns loose.

SHOCK CORD MOUNTS

The shock cord is the line that attaches the nose to the body tube. The term was first used by Orville H. Carlisle, the inventor of the model rocket, because early shock cords were very short and had to absorb the shock of the nose flying off the front of the body tube and being brought up short. The first shock cords were 1/8-inch wide rubber strands identical to those used for rubber-powered model airplanes. This is still used in many kits. But an elastic shock cord isn't required if it's 18 to 24 inches long; stout cotton twine works beautifully and doesn't lose strength or elasticity.

A perfectly good shock cord can therefore be made from an 18-inch length of cotton kite twine. This length allows the nose to slow down sufficiently after it's been popped off the model. Shorter shock cords may break. A ball of cotton kite twine purchased in the hobby shop will provide you with shock cords for a couple of years. Don't use nylon, Dacron®, or other twines

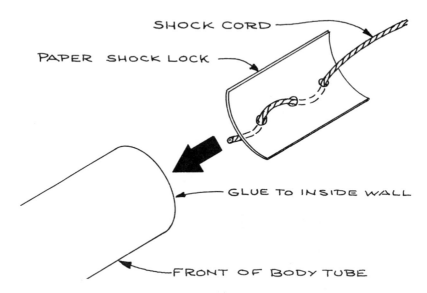

Figure 4-10 How to make and use the shock lock for installing a shock cord in a model rocket body tube.

made from artificial fibers; the heat from the ejection charge gas will melt them.

A shock cord keeps the nose and the rest of the model rocket together during recovery so you don't have to recover two pieces separately; it's usually difficult enough to recover the whole model tied together.

Attaching the shock cord to the body tube has always been one of the difficult problems of model rocket construction. In the early days, we slit the body tube, threaded the elastic shock cord through the slit, and glued everything down flush to the body tube. It took a lot of extra work to hide a shock cord thus attached, especially on scale models. Many other methods of shock cord attachment were tried. In 1969, I developed the paper "shock lock" shown in Figure 4-10. By threading the shock cord through the holes in the shock lock as shown, then gluing the assembly to the inside of the front of the body tube, you don't cut or mar the external surface of the tube or interfere with the recovery package or wadding. This shock lock will usually last for the lifetime of the model. A cotton shock cord gives up from sheer fatigue after fifty or sixty flights, and you'll have to glue in a new shock lock and cord. But if you can keep a model rocket flying for that many flights, you're doing very well.

RECOVERY DEVICES

If you've spent a lot of time building a good model rocket, you'll want to get it back after its first flight in a condition to fly many more times. In addition, a 2-ounce (57-gram) model rocket falling freely out of the sky in a streamlined condition isn't safe at all. Those are two good reasons why you should never fly a model rocket without a recovery device. And it's very simple to make a good recovery device, and model rocket motors are designed to eject such a device in flight.

All the model rocket motors used in single-staged models or in the upper stages of multistaged models are made so that they'll produce a retro-thrust puff of gas at a predetermined time after ignition of the motor. This puff of gas is used to activate or deploy the recovery device. Chapter 5, Model Rocket Motors, explains in detail how this is done. Briefly, the motor puffs a bit of gas that pressurizes the inside of the body tube, pushing the wadding and recovery package forward, dislodging the nose, and permitting the recovery package to exit from the model and deploy in the air.

Many types of recovery devices have been developed and used by model rocketeers. Some of the more successful and common ones are described in detail in Chapter 12, Recovery Devices. For now, we'll take a quick overview of the most widely used ones.

The most common device is a crepe paper or plastic streamer that's tied to the shock cord near the nose end of the cord. When ejected from the model, it streams in the wind and flutters, slowing the model's descent. Streamers are used on small models weighing less than 3 ounces (85 grams). They are most widely used on small high-performance models that ascend to high altitudes or models that are flown from small fields in windy conditions. A model with streamer recovery will return to the ground more rapidly than one with a parachute, and it will not drift as far in a wind.

A streamer is a long, narrow, rectangular strip of crepe paper (preferred for contest work) or thin plastic film. It's usually 1 to 2 inches wide and 12 to 24 inches long. A fairly standard streamer is 1 inch by 18 inches. The best length-to-width ratio is ten to one. A streamer should be made of brightly colored material, preferably bright orange or red, so that it can more easily be seen against the sky, on the ground, or in a tree.

The most obvious recovery device is a parachute. Most model rocket parachutes are made from polyethylene film less than 1/1000 inch thick. Commercially made model rocket parachutes are printed or decorated in bright and contrasting colors and patterns so that they may be more easily seen. A typical model rocket parachute is shown in Figure 4-12. Most model rocket parachutes have six or eight sides with six or eight shroud lines, respectively. The shroud lines are lengths of carpet thread or nylon thread. The

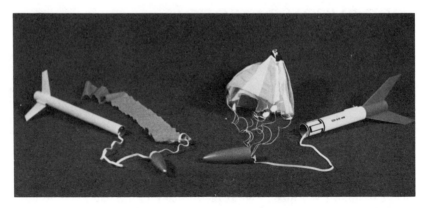

Figure 4-11 Two model rockets, one using a simple streamer for recovery and the other using a plastic parachute for a slower descent.

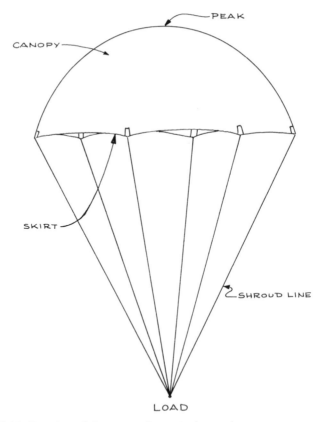

Figure 4-12 Drawing of the parts of a typical parachute.

shroud line length should be at least equal to the diameter of the parachute, and longer if possible. The shroud lines are attached to the parachute by means of tape disks or strips.

Be sure to attach the shroud lines firmly to the edge of the parachute canopy. When a parachute is ejected from a model rocket and opens in flight, it sometimes snaps full of air with a loud pop that can be heard on the ground more than a hundred feet below. This action puts a great deal of strain on a parachute and the shroud lines. If one or more of the shroud lines pulls off, the parachute will lower the model more quickly.

The larger the parachute, the longer and farther the model will drift in the wind. If you put a big parachute into a little model rocket powered by a high-impulse motor, you'll probably never see that model again. It will drift for miles before it lands.

A parachute 8 inches to 10 inches in diameter is usually adequate for models weighing up to 2 ounces (57 grams). Models with 18-inch parachutes have set world duration records. I've lost count of the number of model rockets I've lost with 18-inch parachutes. At the First International Model Rocket Competition in Dubnica-nad-Vahom, Czechoslovakia, in 1966, my parachute duration competition model with an 18-inch parachute was found 17 miles away by a Czech forest ranger who was somewhat astounded to discover a small rocket with American markings in the middle of the Little Tatra Mountains of Europe.

PAINTING

Take the time and trouble to paint your model rocket, even though it may come in a kit with all parts precolored. It will look better if it's painted—if you do a good job. Usually, a model rocket will also perform better if it has a smooth, shiny finish of carefully applied paint.

Model airplane dope isn't highly recommended for painting model rockets although it can be used. The main problem with dope is that it's designed to shrink as it dries; this feature allows it to tighten up the tissue coverings on model airplanes. But on model rockets, this shrinking feature often causes thin balsa fins to warp, or the dope layer pulls away from the body tube where the fins and body tube join. In addition, model airplane dope melts or crazes the surfaces of polystyrene plastic parts.

Hobby stores carry two kinds of paint that can be used on model rockets. These are *enamels* and *acrylics*. Testor's® and Pactra® both make enamels in a variety of colors. However, enamels straight out of the bottle or can are usually too thick, leave a lumpy finish, and change color with age (white gets yellowish). Most model rockets painted with enamels by beginners look pretty bad. I often think that some modelers smear the enamel on with a cotton swab. Acrylic paints made by Floquil® and others can be thinned with water, go on smoothly without showing brush strokes, do not yellow with age, dry to the touch in 30 to 60 minutes, produce a hard, water-resistant finish, and don't smell.

The finest model rocket paint job is one applied by spraying. Enamels, alkyd enamels such as Krylon®, and other kinds of paints are available in a wide variety of colors in aerosol spray cans ranging in price from a dollar up. To use these aerosol spray paints, put your model on a paper wand and give it several thin coats of paint for the best results. Don't try to get a complete, all-covering paint job with one coat; the paint will run, producing an unsightly glob. "Dust" the paint onto the model. You can put on several light coats and have them dry in less time than it takes for one thick, globby, runny coat to dry.

Small hobby-type airbrushes are now available at reasonable cost in hobby shops. These are ideal for painting model rockets. I have several of them and have actually worn out two of them over the years. Water-based acrylic paints are best to use with airbrushes because you can thin the paint with water and clean out the airbrush with water when you change paint colors or get ready to put things away for the night. Pressure for airbrushes usually comes from cans of pressurized carbon dioxide sold in hobby stores; these are all right for quickie jobs, but you'll soon grow tired of buying pressure cans. The next best bet is to buy a small air compressor such as those now available at hobby shops. A compressor will set you back a fair piece of change, but it will quickly pay for itself if you do a lot of spray painting. It's also handy for inflating bicycle tires and athletic equipment, blowing sawdust off things, etc. Some airbrushes will work on the output of a tank-type vacuum cleaner; attach the hose to the end of the vacuum cleaner that blows instead of sucks. In very damp or humid climates or for portable use, you can buy small cylinders of pressurized carbon dioxide which, when used with a pressure regulator to drop the pressure, will produce a steady stream of dry gas for your airbrush.

Paint your model rocket in bright colors that can be easily seen in the air and on the ground. Good colors are fluorescent red or orange as suggested earlier. A model rocket is most highly visible when painted in one solid color with perhaps one black fin. Painting it many colors will tend to camouflage it and make it hard to see.

Fluorescent colors are available in hobby stores in spray cans or in bigger Krylon® cans. Fluorescent colors must be sprayed on the model because painting them with a brush leaves them streaky in appearance. Fluorescent colors must also be applied over a base coat of flat white paint.

The best nonfluorescent colors for maximum visibility depend upon the general sky color and condition in your locale. Against the clear, blue skies

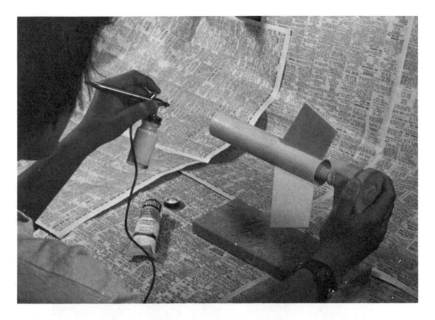

Figure 4-13 A hobby airbrush provides a beautiful, even coat of paint and is well worth the expense. Be sure to use lots of newspaper to prevent the overspray from getting everywhere.

of the American Southwest, the best colors are white, orange, or red. Against the cloudy, gray, hazy skies of the Midwest, South, or East Coast, dark colors such as maroon or black work well. Some silver paints show up well against cloudy or gray skies, but they often wash out by taking on the color of the sky itself; then they tend to disappear against the sky. Greens, browns, and sky-blues blend too well with the surrounding environment. One not-too-bright model rocketeer once painted his rocket in accurate military camouflage colors that really worked; he couldn't find his model after it landed on the ground.

DECALS AND TRIM

Most competition models don't carry much decoration other than a decal or marking showing the contestant's sporting license number as required by national rules. Competition modelers have learned that putting decals and strongly contrasting paint patterns on a model tends to make it more difficult for the human eye to see and follow. Note that military camouflage and the natural coats of animals use a number of contrasting colors in random patterns.

However, the application of decals and trim to sporting models results in a striking change in the appearance. The smooth, streamlined, uncomplicated shapes and lines of the model suddenly seem to come alive; the model begins to look more like its big brothers at the Cape. To obtain a good-looking model that resembles a real rocket, it's important to know what kind of decals and trim to use. This is because color patterns and markings on the real ones have definite functions.

Figure 4-14 The proper placement of a few simple decales on a model rocket will greatly improve its appearance.

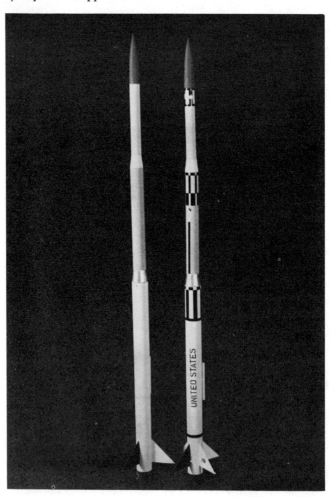

The most important decal or marking is the roll pattern of regular stripes, checks, or other marks around the body. At the Cape, roll patterns are painted on rockets for photographic data purposes so that engineers can determine from camera films how the rocket rolled or tilted in flight. If you roll one of your models in your hand, you'll see how the roll pattern decal changes appearance and how it's possible to determine roll rate from that change in appearance.

Usually one fin is painted black or some contrasting color. This is to provide a roll reference point. Fins sometimes have different paint patterns on them to identify them; engineers looking at films of the flight can tell which fin they're looking at.

Most NASA space vehicles and sounding rockets carry a painting of the American flag and the words "United States" or "USA" on the body of the first stage. The star-and-bar military insignia is never applied to the nonmilitary NASA rockets. That insignia, however, and the words "U.S. Air Force" or "USAF" usually are placed somewhere on the body of an Air Force rocket. Often USAF rockets also have some sort of squadron or command emblem, or other marking on them.

Numbers are usually applied to fins but may also appear on the body. Numbers and letters are usually applied in an orientation that permits them to be read when the rocket is in the launching or flight attitude.

Follow the instructions on the back of the decal sheet for applying decals. Always blot a water-applied decal, especially large roll patterns, to get the air and water bubbles out from under them. A water-applied decal should be allowed to dry for an hour or more. Otherwise, it may rub off the model.

Some decals are now vinyl-backed and self-adhering. This means you've got to get them on right on the very first try.

Other markings may be applied to a model using an India ink pen such as the Mars-700® or Rapidograph®.

Kits and supplies for making your own decals are available in some hobby shops. Check in the arts and crafts department. Or check in an art supply store.

After all decals and markings have been applied and are dry, apply one light coat of dull or gloss transparent spray over the entire model, depending on whether you want it to have a flat or glossy finish. Dulling spray will kill the shine of the decals and make the model look bigger.

STORAGE AND CARRYING

Although your model rocket may be strongly constructed and capable of flying through the air at half the speed of sound or better, it's surprisingly easy to break when it's on the ground. Little Brother may get to it. Fido may decide it's a superior and very tasty new kind of bone. Big Clod may step on it. Fumblefingers may drop it three feet to a concrete floor where it will shatter into ten thousand little bitty pieces, most of which you'll never be able to find. Or it simply may get trashed by an overzealous Mom who doesn't understand that you're trying to get a hundred flights on that ratty old model.

Put your models away between flight sessions. Make a display rack for them. Or hang them from the ceiling by threads. Or put them in a big drawer—but be careful not to shut the drawer on the fins. A big cardboard box is great for storing model rockets.

To keep a model rocket from being damaged in storage and during transit to the flying field, put it in a plastic bag. The air trapped inside the bag with the model acts like a cushion to protect the model. I've carried model rockets back and forth in airplanes across the Atlantic Ocean in plastic bags inside a cardboard box without the slightest damage or paint scratch. I always open the plastic bag of a kit model very carefully because I save that bag to put the completed model in.

Now that we've covered some of the basics of model rocketry, we can talk about some of the more advanced parts of the hobby. Before we go on, however, reread these first four chapters again just to be sure you understand what we've been talking about and won't forget it. Even after you've read about some of the more complex aspects of model rocketry, come back and review these fundamentals from time to time. That way, you'll have a sound understanding of these fundamentals and won't be held back by making basic mistakes.

5

Model Rocket Motors

The device that makes model rocketry a hobby and a sport rather than a disaster is the model rocket motor.

Some people call it a model rocket "engine," but there is a subtle distinction between a "motor" and an "engine." A motor is defined as "something that imparts motion," while an engine is defined as "a machine that converts energy into mechanical motion." Thus, a steam engine or an internal combustion engine is definitely an engine. A model rocket motor is truly a motor. Technically, it's a small reaction device for converting the energy of high-temperature, high-pressure gas into motive power without the use of gears, cams, linkages, pistons, turbines, etc.

There are many kinds of rocket motors and engines. They're usually categorized by the type of propellant they use—liquid propellant or solid propellant—or by some unique form of energy they convert into motion— nuclear rockets, ion rockets, nuclear pulse rockets, etc.

All model rocket motors are solid-propellant rocket motors. There's a reason for this as we'll find out.

A solid-propellant model rocket motor is an inexpensive, highly reliable package of power that comes all ready for use. It will provide both the propulsive force to thrust a model rocket hundreds of feet into the air and the means for ejecting the recovery device.

Model rocket motors are the world's most reliable rocket motors. By 1981, almost 200 million of them had been manufactured and used without producing any severe fire hazard or personal injuries more serious than an occasional burned finger.

A model rocket motor appears to be a very simple device, but it's actually very complicated. Making one costs a large amount of money and requires extensive and very special equipment, plus a lot of knowledge. Strict safety precautions and expensive safety equipment must be used in making model rocket motors. This explains why model rocketeers leave the making of rocket motors to professionals, the model rocket manufacturers.

There is no safe way to make a rocket motor of any type. This is a statement of fact, not a matter of opinion. There is no way you can make a rocket motor that will be as inexpensive and as reliable as a commercial model rocket motor. A model rocket motor is a factory-made device that is subject to rigid quality standards, quality controls, and statistical batch sampling and testing procedures. It's very reliable and will do exactly what it's designed to do. You, as a model rocketeer, don't have to handle dangerous chemicals, worry about whether or not the motor will have proper thrust, or take extensive and expensive safety precautions.

Never forget that a model rocket motor is not a toy. You must understand this right from the start. Safe as it has been proven to be, if you don't use it correctly and in accordance with instructions, it can hurt you. This is also true of most technical devices in the world around us.

A model rocket motor is packaged reaction propulsion power for models and should be used for no other purpose.

Model rocket motors should be your introduction to the fact that technology will work for you if you handle it properly—and that it can bite you if you don't. Cro-Magnon people learned about technology the hard way when they got burned by their cave fire. Fortunately, you don't have to learn everything in that difficult school of hard knocks called experience, but you must be willing to listen to people who know something about it—like me— and follow their instructions and teachings. Soon, you'll know more than your teacher.

As of this writing, there are 175 different types and makes of model rocket motors available in the United States that have been certified by the Standards and Testing Committee of the National Association of Rocketry (NAR) as meeting the reliability and performance standards of the NAR, the Hobby Industry Association, and the National Fire Protection Association. This is the greatest selection available in any country in the world. From this broad selection, you should be able to choose a model rocket motor that meets your requirements.

A typical solid-propellant model rocket motor is shown in Figure 5-2 as though it were cut down the middle to expose the innards. (Don't do it! It's safer to look at the drawing!)

The motor *casing* is made from tightly wound paper with carefully controlled dimensions or from special composite plastic. Most model rocket motors use paper because it's strong, fire-resistant, and doesn't conduct heat easily. Advanced high-performance model rocket motors often use composite plastic casings for even greater strength and heat resistance. The dimensions of model rocket motor casings vary according to the power, manufacturer, etc. However, there are some basic model rocket motor sizes, as shown in Table 4.

The *nozzle* is made from ceramic formed into a carefully designed size and shape. You should never try to alter the nozzle because the slightest change in dimension of the nozzle throat—as little as a thousandth of an inch—could drastically alter the operation of the motor.

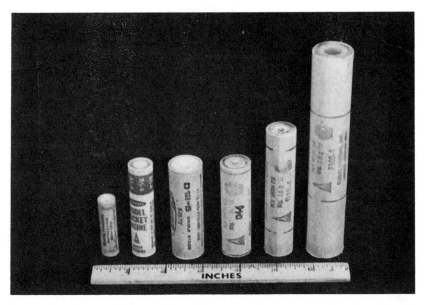

Figure 5-1 Some of the model rocket motors manufactured in the United States at the time of this writing. Units are representative of some of the different sizes available.

Figure 5-2 Cross-section diagram of a typical solid propellant model rocket motor.

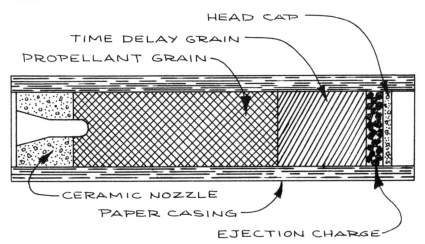

TABLE 4
Model Rocket Motor Size Chart

Correct and complete as of August 1982

Family or Series	Manufacturers	Diameter × Length (millimeters)	(inches)
"Mini"	Estes & Centuri	13 × 45	0.50 × 1.75
"Standard"	Estes, Centuri, & FSI	18 × 70	0.69 × 2.75
Type D	Estes & Centuri	24 × 70	0.945 × 2.75
FSI D18, D20, E5	FSI	21 × 95	0.83 × 3.75
FSI E60	FSI	27 × 114	1.06 × 4.50
FSI F	FSI	27 × 150	1.06 × 5.91
CRT E30	CRT	29 × 66	1.14 × 2.60
CRT E45	CRT	24 × 83	0.945 × 3.27
CRT F50	CRT	29 × 83	1.14 × 3.27
CRT F67	CRT	29 × 102	1.14 × 4.02

NOTE: FSI = Flight Systems, Inc.
CRT = Crown Rocket Technology

The *solid propellant* is a hard piece of combustible chemical material with controlled burning characteristics. To a model rocketeer, the composition of this solid propellant isn't important because he's far more interested in the performance of the motor. However, some of the high-performance large model rocket motors use composite solid propellants derived from those developed over the past 25 years for space rockets.

Once ignited, the kind of solid-propellant motor shown in Figure 5-2 burns forward from the nozzle, producing more than two thousand times its solid volume in hot gas. This gas shoots out the nozzle and produces thrust in accordance with Newton's ThirdLaw of Motion. More about this later.

Ahead of the solid propellant is the *time delay charge*. This is a piece of very slow-burning solid propellant that produces very little gas and therefore practically zero thrust. Its action allows the model time to coast upward on its momentum to apogee. If there were no time delay charge in the motor, the ejection charge would deploy the recovery device from the model at a low altitude and a high speed. The model and its recovery device aren't strong enough to withstand this sort of flight behavior even once.

The time delay charge lasts for several seconds, depending upon the type of motor you install in the model before flight. The end of burning of the time delay charge automatically activates the *ejection charge*. This produces a quick puff of gas that pressurizes the inside of the model rocket body tube, pushes the recovery wadding and package forward, dislodges the nose, and expels the recovery device from the model. The ejection charge is held in place in the motor with a *head cap* that is either a paper cap or a thin ceramic plug that's shattered when the ejection charge is activated.

And that's all there is to it.

And it works.

You should never attempt to reload a used solid-propellant model rocket motor casing. It's been designed for only a single use. If you want to cut apart a *used*, expended motor casing (never a loaded one), you'll see that the inner surface of the casing has been charred and ablated away during operation in much the same manner that the old space capsules' heat shields were ablated by atmospheric frictional heat when they returned to earth. This casing ablation weakens the casing to the point where it probably isn't safe to use it again. In addition, as I pointed out earlier, the handling of rocket propellants and the making of rocket motors are jobs for trained experts—and you aren't one.

You'll find a lot of information printed on a model rocket motor casing and even more in the package and instructions that come with the motor. Read these! And check for the statement, "NAR Certified." This means that the Standards and Testing Committee of the NAR, the nonprofit spokesman for model rocketry in the United States, has tested samples of that type and make of motor and has determined that the motor type meets or exceeds the strict set of performance and reliability standards jointly developed by everyday model rocketeers like you and me who are members of the NAR and by the model rocket motor manufacturers who are members of the Hobby Industry Association (HIA). If a model rocket motor or its package or instructions doesn't have "NAR Certified" printed on it, you should be wary of it. In some states, it's against the law to sell or use a model rocket motor that doesn't have the Safety Certification of the NAR.

All NAR-certified model rocket motors carry on their casings the universal United States model rocket motor code that tells what kind of motor it is and how it will perform. This NAR motor code is simple. It consists of a letter, a number, a dash, and a final number.

A typical example might be: B4-6.

The *first letter* of the code indicates the power range of the motor.

How do you figure the power of a model rocket motor? In terms of horsepower? Starpower, maybe? No, in terms of *total impulse*, which is a factor derived by multiplying the average thrust by the thrust duration. Or, more accurately, total impulse is the area under the thrust–time curve.

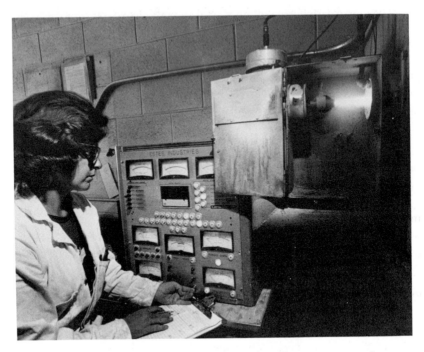

Figure 5-3 Static testing a model rocket motor to determine its performance. Manufacturers have developed complex electronic testing equipment to insure high product reliability. (Estes Industries, Inc.)

Sound confusing? Well, it won't be if you take it one step at a time.

The *thrust* of a motor is the amount of force, or push, produced when it's operating. The jet of supersonic gas rushing out of the motor nozzle produces a force according to Sir Isaac Newton's Third Law of Motion. Stated simply in words, this is: For every acting force, there is an equal and oppositely directed reacting force. Written as an equation, the universal shorthand of science and technology, this is: $MA = ma$.

The thrust of a model rocket motor is rarely constant. It changes with time. Therefore, we must further define which thrust level we're speaking of. There are several. *Maximum thrust* is the highest amount of force produced by the motor during operation, regardless of when that occurs during the period of operation. *Average thrust* will be defined in a moment.

Because model rocketry is an international sport, motor performance specifications are in the units of the international metric system—more specifically, in terms of the meter–kilogram–second (MKS) system. Motor thrust is therefore given in terms of *newtons*, named after the aforementioned Sir Isaac Newton. A newton is defined as the force required to accelerate 1 kilogram (2.2 pounds) of mass at a rate of 1 meter per second per second (3.28 feet per second per second). It's easy to convert newtons of force into pounds of force:

$$1 \text{ pound of force} = 4.45 \text{ newtons}$$

The model rocket motor begins to produce thrust at the instant *ignition* of the propellant occurs. *Burnout* is the instant that the motor ceases to produce measurable thrust. The length of time that thrust is produced is called *duration*, and it's measured in seconds of time. A duration of 2 seconds is very long for a model rocket motor. The time interval from ignition to maximum thrust is known as *T–max*. This is an important parameter to know if your model rocket is heavy; it will help you choose the right launch rod length to use, as we'll see later.

To accurately determine thrust, maximum thrust, duration, *T–max*, and other performance characteristics, a model rocket motor must be given a static test. It's fastened into a fixture on the ground with measuring and recording devices attached, then is ignited and operated.

The data record of a static test produces a thrust–time curve such as the generalized one shown in Figure 5-4. This is a typical thrust–time curve for most model rocket motors. Notice how the thrust rises rapidly after ignition to a high maximum thrust. This accelerates the model rapidly to high speed, ensuring that it has sufficient airspeed for the fins to stabilize it when it leaves the launch rod. Then the thrust level settles back to a lower value, a sustaining thrust that accelerates the model to higher speeds during the climb.

When we static test a model rocket motor, it's no longer usable for flight. How do we know that other motors of its type are going to produce the same performance? Answer: We don't know for certain, but we can infer that they will. This inference is based upon confidence gained from testing a large number of motors of the same type and is based upon statistical sampling and proven testing techniques. You can delve into this area of mathematics in detail if you're a math shark—and a lot of model rocketeers are. Model rocket motors follow the laws of statistical difference just like everything else.

Each model rocket manufacturer has his own high-precision model rocket motor static test stand. Today they're a far cry from the simple equipment using springs and rotating drums that were common in the early days and are still common in many high schools today. However, solid-state electronics has made static testing easier and has brought the cost and complexity of a static test stand within the means and capabilities of high school model rocket clubs.

Manufacturers subject a random sample—usually 2 percent—of each motor production batch to static tests to check performance and conformance to NAR standards. If the test sample fails, the entire production lot must

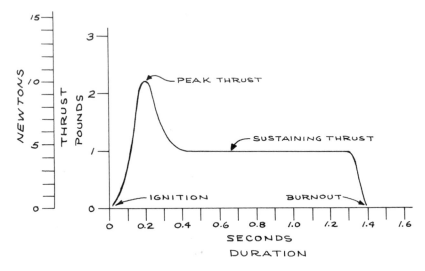

Figure 5-4 Typical thrust–time curve of an end-burning solid propellant model rocket motor showing various events.

either be rejected and destroyed or corrected so that test samples perform properly. Over a period of time, a manufacturer gets much information from thousands of static tests. He can have a high degree of confidence that a small random sample of a lot represents the rest of the lot, that the rest of the motors will perform the same as the test sample. This is a fascinating area of science that few people are aware of, and statistical quality sample tests of products are a well established area applied to everything from ball-point pens to semiconductor computer chips.

After all, if 200 million model rocket motors have been made, it means that 2 million static tests have been made. This is a *lot* of data! It provides a very large statistical universe.

In addition, the NAR Standards and Testing Committee obtains and tests random samples of model rocket motors from all sources. Committee members check to see that production motors sold in stores are the same as the original test motors submitted by the manufacturers for certification. This double-checks the manufacturers' own quality standards programs. And all certified motor types are tested for *recertification* by the NAR every three years.

Yes, there have been some motors from some manufacturers that haven't made the grade and have been denied certification. And there have been some types that have had their certifications withdrawn. Who, what, why, and when are of no importance. But it does prove the basic NAR philosophy, backed by the National Fire Protection Association: model rocket safety is a combination of strict product quality control *plus* your own good common sense in using the product safely. For almost a quarter of a century, it's worked. A lot of people didn't think it would, but the manufacturers, the NAR, and millions of model rocketeers proved it would and does. That's something to be proud of, and I'm proud of you for doing it.

As stated earlier, thrust and time, or duration, are equally important in determining motor performance. It does no good whatsoever to have a thrust of 100 pounds (445 newtons) if it lasts only 0.01 second! Likewise, a thrust of only 0.1 pound (0.445 newtons) lasting as long as 10 seconds probably won't even lift the model into the air. The performance factors of a model rocket motor that really determine the flight characteristics of a model rocket are thrust, duration, and their combination, *total impulse.*

It was stated above that, roughly speaking, total impulse is determined by multiplying thrust by duration. However, thrust isn't constant, so we usually determine total impulse by a more precise method: measuring the area under the thrust–time curve, which amounts to the same thing. For math buffs, we integrate the area under the thrust–time curve.

Using a thrust–time curve generated during a static test, we can determine this area by laying the recorded chart down over a sheet of quad-ruled paper and counting the number of squares inside the thrust–time curve. A far more accurate piece of data can be generated by using a plane polar planimeter to do the same thing. Most manufacturers use static test stands that perform this integration electronically to produce a digital result.

The result comes out in terms of thrust × duration—newtons times seconds or newton–seconds. In the old English system, which is not easy to use in model rocket performance calculations, this would be in terms of pound–seconds.

Total impulse roughly determines how fast and how high a model rocket will fly. Flight calculations discussed in detail in a later chapter will show why this is so. For now, it's enough to know that a motor with higher total impulse will take a model rocket to a higher altitude.

This is why the NAR rates model rocket motors in classes based on their total impulse. The NAR total impulse classifications are shown in Table 5. All American manufacturers use these classifications.

Thus, the first letter in the NAR motor code tells you the total impulse range of the motor. Most motors perform 10 percent below the top of the range to allow for statistical variation, although some of the larger motors may fall in the middle of the range. However, you don't have to know the precise total impulse of a motor to get some idea of its general performance because each total impulse class is roughly double that of the previous class. This means that a Type B motor will have twice the total impulse of a Type A, and that a Type C will have twice the total impulse of a Type B and four times that of a Type A.

TABLE 5
Model Rocket Motor Total Impulse Ranges

FAI (international) and NAR (USA) standards.

Type	Total Impulse (newton–seconds)	Total Impulse (pound–seconds)
1/4A	0.000 to 0.625	0.000 to 0.14
1/2A	0.626 to 1.25	0.15 to 0.28
A	1.26 to 2.50	0.29 to 0.56
B	2.51 to 5.00	0.57 to 1.12
C	5.01 to 10.00	1.13 to 2.24
D	10.01 to 20.00	2.25 to 4.48
E	20.01 to 40.00	4.49 to 8.96
F	40.01 to 80.00	8.97 to 17.92

The *first number* in the NAR motor code (Type B4-6 in our example) tells you the *average thrust* of the motor in newtons. This is a derived, or calculated number. It is determined by dividing the total impulse by the duration. It indicates what the thrust *would be* if it were *constant* from ignition to burnout. Average thrust is a useful piece of information for altitude prediction. It also tells you how fast your model will accelerate into the air and roughly how fast it will be going when it leaves the launch rod.

We can clarify this by looking at Figure 5-5, 5-6, and 5-7.

Figure 5-5 shows the *idealized* thrust–time curve of a *constant–thrust* hypothetical Type B10 motor with an average thrust of 10 newtons for a duration of 0.5 second. This is a high-acceleration motor useful for flying heavy models or for flying regular models in high winds to prevent weathercocking; more about this later.

The hypothetical Type B5 motor whose thrust–time curve is shown in Figure 5-6 has an average thrust of 5 newtons and a duration of 1.0 second. The Type B5 motor has exactly the same total impulse as the Type B10. But the Type B5 would be a better choice for a small, lightweight high-altitude model.

Both the Type B5 and Type B10 have the same total impulse—5 newton-seconds (abbreviated N–sec). But their thrust characteristics are quite different, and they would make the same model perform quite differently.

Figure 5-7 shows the thrust–time curves of two motors with identical average thrusts but different total impulses. The Type A5 has 5 newtons of

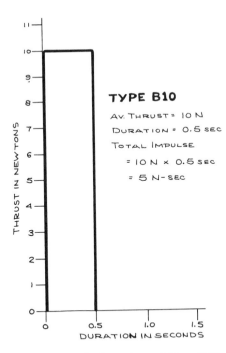

Figure 5-5 Thrust–time curve of a hypothetical Type B10 motor.

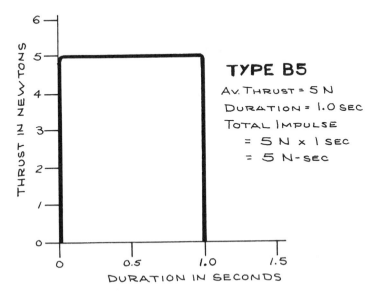

Figure 5-6 Thrust–time curve of a hypothetical Type B5 motor.

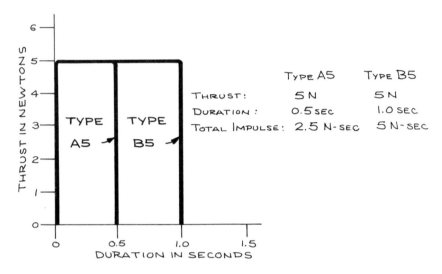

Figure 5-7 Thrust–time curves of hypothetical Type A5 and Type B5 motors showing equal thrusts but different total impulses.

thrust and a duration of 0.5 second. The Type B5 also has 5 newtons of thrust, but a duration of 1.0 second. Which motor type will take the same model to a higher altitude? Obviously, the Type B5 will, because thrust will be applied to the model for a longer period of time.

You can get a rough idea of the motor's duration by dividing the maximum total impulse of its class by its average thrust number.

The *number following the dash* in the motor code tells you the number of seconds after burnout before the motor activates the ejection charge and deploys the recovery device.

Thus, in our example, the Type B4-6 motor has a total impulse range of 2.51 to 5.00 N–sec (probably being 4.90 N–sec at the top of the range), an average thrust of 4 newtons (little less than a pound of thrust), and a time delay that will activate the recovery device 6 seconds after burnout.

This is nearly everything you need to know in order to select the proper motor for your model rocket. You can't get by with any less information than what's in the NAR motor code. The system's been in use since 1960. There have been many attempts to simplify it, but nobody yet has been able to devise a coding system that's so easy to remember but still tells all the basic information you need to know about a motor. If anybody says the system's too complicated for him to learn, it's my guess he probably doesn't have enough gray matter to get involved safely in model rocketry in the first place.

You must always match the motor to the model *before* the flight. You must select a motor with enough average thrust to accelerate your model safely into the air; the heavier the model, the higher the average thrust should be. Most importantly, you must select the proper time delay. In the beginning, *follow the recommendations of the kit manufacturer regarding time delay.* Later, you'll have enough experience to decide for yourself which time delay to use.

If the time delay is too short, the model will deploy its recovery device while it's still climbing. This will probably happen at high airspeed, and the aerodynamic forces may tear the recovery device to shreds or rip it entirely off the model.

If the time delay is too long, the model will climb to apogee, pause, arc over, and begin to fall back toward the ground. It will gain speed as it falls. Again, the recovery device may be deployed at high speed and tear itself to pieces; the model itself may also be destroyed in the process. These "cliff-hanger" flights are no fun, and they could be hazardous.

RULE OF THUMB: When in doubt, use a *shorter* time delay. If the recovery device deploys while the model is still going up, use a longer time delay motor on the next flight.

Some model rocket motors are dash-zero types, i.e., the motor code ends in a zero; Type B6-0, for example. These motors have no time delay or ejection charge. They are intended for use in the lower stages of multistaged models. See Chapter 11, Multistaged Model Rockets.

ANOTHER RULE OF THUMB: Don't overpower your model. First flights should be made with the lowest total impulse motor type recommended by the kit manufacturer. If you overpower a model, you're likely to lose it on the first flight. The kit manufacturer and the hobby store will love you because you'll have to buy another kit. The motor manufacturer will love you, too, because the higher-powered motors have more propellant in them and so cost more money.

Without mathematics and computer programs, you can estimate how high your model will go. As we'll see in a later chapter, it's possible to predict with great accuracy the altitude a model will achieve, but that's a bit more advanced. For starters, you can expect the following performance from a basic beginner's model weighing about 1 ounce (28 grams), using a 3/4-inch (20 millimeter) body tube, and equipped with these motors:

Type A8-3: 500 feet (150 meters)
Type B6-4: 1,000 feet (300 meters)
Type C6-5: 1,600 feet (500 meters)

The time delays shown in these motor codes are about right for a well-built, streamlined, 1-ounce model rocket.

Thus far, we've discussed only the smaller types of model rocket motors that you're likely to be using when you get started. Most model rocketeers fly with motors ranging from Type 1/2A to Type D. All of these, with one

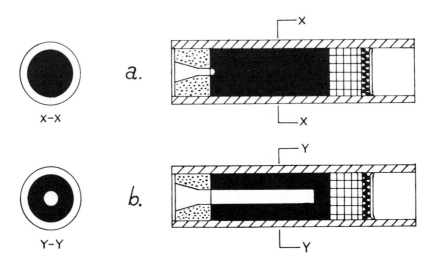

Figure 5-8 Cross sections of hypothetical model rocket motors showing (a) typical end-burning propellant grain, and (b) typical core-burning propellant grain.

or two exceptions, look like the cutaway drawing of Figure 5-2 and produce a thrust–time curve similar in most respects to Figure 5-4. This type of motor is categorized in rocket propulsion terminology as a *restricted end-burning grain* design. Some propulsion engineers call it a "cigarette burning grain" because burning starts on one end and progresses to the other as a cigarette does.

A solid rocket propellant burns only on its exposed surface and produces hot gas in proportion to the amount of propellant area that's burning at any given instant. To get more gas and therefore more thrust, a rocket engineer will design the propellant grain—as the propellant charge of a solid-propellant rocket engine's called—so that more burning area is exposed.

There are a number of different grain configurations used in professional solid-propellant rocket motors. For example, the Space Shuttle Solid Rocket Boosters (SRBs) have a hole up their middle that's star-shaped; it also has a different shape at different locations in the motor to cause more or less propellant area to be exposed for burning at any given time during operation, thus effectively "throttling" the SRB thrust, keeping it at a level that will provide a comfortable ride for the people in the Orbiter.

There are two basic grain configurations used in model rocket motors. The first, which we've discussed, is the restricted end-burner. The other, used for high-impulse large motors in Type E and Type F categories and in a couple of small high-thrust motors (Estes A10-OT, A10-3T, B8-0, B8-5,

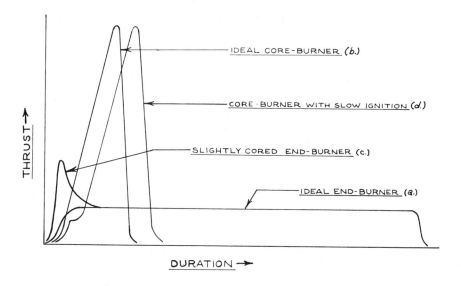

Figure 5-9 Thrust–time curves of various model rocket motor types with differing propellant grain configurations.

and B8-7, as well as the Centuri "Super-C" Type C5-0 and C5-3), is the *unrestricted core-burning* grain configuration. These two are shown in Figure 5-8.

Since the end-burner has more or less a constant burning area exposed at all times during operation (except right at the start where the grain is "dimpled" at the nozzle to provide extra burning area that in turn produces the thrust spike noted in the thrust–time curve), the end-burner produces almost constant thrust.

With its hole up its middle, the unrestricted core-burner grain has far more propellant area exposed for combustion. With this greater burning area available, this grain design produces more thrust. But it burns not from one end to the other, but *outward* from the center. The thrust *increases* during burning as more propellant area is exposed. Also, the duration of a core-burner is less.

The thrust–time curves for various types of grain configurations is shown in Figure 5-9. Curve A is that of a classical end-burner. Curve B is that of the typical core-burner. Note that the core-burner's thrust starts out low and increases as its center core burns and increases in diameter, thereby increasing burning surface. Curve B is not especially good for model rockets which need a short kick in the tail to get them moving quickly off the launch rod and into the air at a speed that will allow their fins to maintain stability.

Therefore, the advanced core-burner model rocket motors use modifications of the simple hole-in-the-middle. They are built so that they have slightly more propellant area available near the nozzle. Or they may contain two different propellant types with different burning rates. Some core-burners achieve their high initial thrust by virtue of the way they're ignited.

For example, if you place the igniter in the nozzle of a core-burning motor, you'll get a slow ignition and a slow thrust buildup. If you place it at the forward end of the core, the thrust buildup will be much faster. *Ignition of core-burning high-performance motors is critical!* Be sure to use the igniter specified by or supplied by the manufacturer of the motor, and be absolutely sure that you install the igniter exactly as the manufacturer's instructions tell you.

Because these advanced, high-performance, core-burning model rocket engines are more difficult and critical to handle, get a lot of experience with motors up through Type D before tackling the Type E and Type F motors.

As you can now see, a model rocket motor, although appearing to be a simple device, is really somewhat complex. It's the result of decades of development by both professional rocket engineers and model rocket motor manufacturers. Each motor type took months of work to bring it up to the level of performance and reliability that would permit it to be NAR-certified and sold to the public. Many thousands of dollars were required to make the special equipment necessary to manufacture such reliable motors. Often this equipment is hazardous to operate. A great deal of credit must be given to the model rocket motor manufacturers who undertake grave risks to make that simple motor. It's packaged Space Age power. Treat it properly, and it will perform with great reliability.

6

Ignition and Ignition Systems

To fly a model rocket properly and safely, you must have an electric ignition system and a launching pad.

Today, all model rocket manufacturers sell one or more electric ignition systems and launching pads either in ready-to-use or kit form. I highly recommend that you purchase these products, although you may be perfectly capable of making them yourself. The manufactured units contain all the necessary and desirable safety and operational features. They may actually cost less than if you purchased all the parts separately to build your own designs.

An ignition system and launch pad are what are known in professional rocketry as ground-support equipment, or GSE. They are also called capital equipment items because, unlike a model rocket motor, they're used over and over again for many model rocket launchings. So, even though good GSE can cost you ten dollars or more, relax in the knowledge that it's a one-time expense. You can use that ignition system and launch pad for flying thousands of models over several years if you get good equipment to start with and maintain it properly.

I'm using a launch pad I got in 1969 and an ignition system I built in 1962. True, I've gone through several different types of 12-volt batteries since then, but one must anticipate replacing a rechargeable battery every three to five years.

You may not like or believe some of the things I talk about in this chapter. However, I've learned the hard way in some cases, and I've been trained in the safety and handling of explosives, pyrotechnics, rocket motors, fuses, ignition systems, and other subjects starting decades ago at White Sands Proving Grounds (now White Sands Missile Range) in New Mexico. Some of my teachers had *really* learned the hard way; they'd lost some parts like fingers. I've still got all my fingers, two eyes, and the hair on my head. If you do things right and listen to people who know, you, too, can proceed through life with all of your parts and faculties intact.

To help you understand GSE, I'll describe how to make a simple electrical ignition system and an inexpensive launch pad. I'll even describe more complex GSE. However, keep in mind that you're probably better off buying commercial prefabricated kits.

This chapter will cover ignition and ignition systems—how to get the model rocket motor ignited in the first place. The next chapter will be devoted to launchers and launching techniques.

When they launch the Space Shuttle at Cape Canaveral, they don't light a fuse and run away. They do it electrically by pushing a button.

Model rockets are also ignited by remote electrical means. Electrical ignition is simple, reliable, and safe. It's not only illegal to attempt to ignite a model rocket with a fuse in nearly all states, but also very dangerous. You must not do it!

So that you'll completely understand why I'm so strongly adamant about this point, here are some of the reasons why you must not use fuses:

1. Fuse ignition is not reliable. Some kinds of fuse will not fit into the nozzle of a model rocket motor and therefore will not ignite the motor. Some large model rocket motors of the core-burning type cannot be safely ignited except with the electrical igniter especially designed for that motor. It's also possible to have a fuse suffer a hangfire, to borrow an old gunnery term. In this case, remnants of the fuse may smolder for more than 30 minutes up in the nozzle where you cannot see; you don't know when, how, or if ignition is going to take place. And you don't dare go to the model to find out!

2. Fuses cannot be timed. The sort of fuse that people can sometimes buy has burning rates that aren't reliable. The fuse could flare up in a fraction of a second, and the model could take off in your face. Never, never, *never* trust the burning rate of a fuse—not when your safety and that of other people depend on it.

3. Fuse ignition gives you absolutely no control over the launching instant. You cannot stop the lift-off in the last split second if you have to—not with a sputtering fuse out on the launch pad under the model. I've actually seen the following events take place during the last 5 seconds of model rocket countdowns: a) a low-flying airplane (that shouldn't have been there anyway) appeared over the crest of a hill and flew directly over the launch site at an altitude of about a hundred feet; b) somebody got excited and ran out to the launch pad, completely ignoring the shouts and screams of the Range Safety Officer; c) a strong gust of wind blew the launch pad over so that the model was pointed directly at me, d) an unsupervised child spectator, just a baby, ran into the launch area chasing a ball, and 3) the yo-yo hooking his model up on the next launch pad didn't pay attention to what was going on and backed away from his own launcher directly into the launcher and model that was in the final countdown

stages. Because we were flying by the rules and using electric ignition, the countdowns were stopped immediately in safety. They could not have been stopped if a fuse had been sputtering out on the launch pad.

4. With fuse ignition, glowing, hot remnants of the fuse are ejected from the motor nozzle upon ignition. These can fall into dry grass and other flammable material around the launcher, starting a fire. Fuses have started nearly all the grass fires reportedly caused by model rockets. In several cases, a sputtering fuse has fallen out of the motor nozzle, landed in dry grass around the launch pad, and started a dandy little grass fire that required the fire department to put it out. In general, fire marshals and fire chiefs are now friendly to model rocketry when conducted according to the rules, but they don't like fuses. Who can blame them?

5. At Cape Canaveral, they do not believe it's safe to ignite the Space Shuttle with a fuse. When professional rocketeers go back to lighting fuses with matches and running for the safety of the blockhouse, we model rocketeers *might* reconsider our requirement for electric ignition—maybe.

6. Fuses are for fireworks, not model rockets.

Simple, reliable, and safe electric ignition gives you the opportunity to stand at a safe distance from your model and send it on its way with a professional countdown that gives you complete control over the exact moment of launching. You can stop the countdown up to the instant your finger comes down on the launch button. If there's a misfire—and we've all had them, even with the most careful preparation—you can disarm the electrical circuit with better than 99 percent confidence that the model isn't going to ignite and lift off.

Every model rocket motor sold today comes with an electrical igniter and with explicit, complete instructions on how to use that igniter properly. Although I'll discuss electrical ignition in general here, you must *always read and follow* what each manufacturer says about the electric ignition of his type of model rocket motor. Not all model rocket motors can be safely and properly ignited using the general methods described herein. Often, special igniters and methods must be used to ensure fast, reliable motor ignition. This is especially true of the larger core-burning Type E and Type F model rocket motors available today. It is also true of *all* the model rocket motors presently manufactured by Canaroc, Ltd., and imported into the United States from Canada.

All electrical igniters are based upon the principle of the electrical heating element. Nearly all of them are hot-wire igniters or are started by hot wires. The basic element is a short piece of very fine nickle–chromium wire called "nichrome" wire. This is the same wire that's used in an electric toaster. And it works the same way. When electric current is passed through it, its electrical resistance and inability to pass the current causes it to get hot. The nickle and chromium from which it is made won't melt or lose strength at these

Figure 6-1 A model rocket is always launched by controlled electrical means using a launch pad and an electrical ignition system. (Estes Industries, Inc.)

red-hot temperatures, as would iron or copper. This hot nichrome wire, in contact with the propellant, then either raises it to the ignition temperature or activates other materials in the igniter itself. An igniter that uses a hot nichrome wire to initiate other materials which flare up and in turn ignite the propellant is called a "squib" igniter, or simply a "squib."

The ignition temperature for most model rocket motor propellants is in excess of 550°F. Some high-performance model rocket motor propellants also

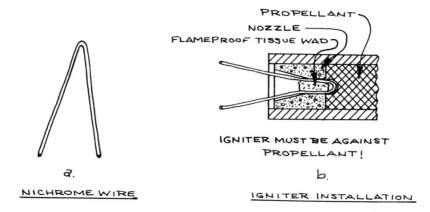

PROPELLANT
NOZZLE
FLAMEPROOF TISSUE WAD

IGNITER MUST BE AGAINST
PROPELLANT!

a.

b.

NICHROME WIRE

IGNITER INSTALLATION

Figure 6-2 Drawing of a typical nichrome hot-wire igniter and its installation in a model rocket motor. This igniter type requires much more current and a bigger battery than some low-current igniters now available.

require both temperature and pressure to ignite, and thus a squib igniter produces both a flare of flame and a brief pulse of pressurized gas.

Once the propellant has been ignited on its surface, it continues to burn on its surface until all of it is consumed.

A typical hot wire igniter is shown in Figure 6-2. It will work in most Estes and Centuri model rocket motors from Type 1/2A to Type D. A 2-inch length of #32 nichrome wire is doubled over and inserted up the nozzle until it's in contact with the propellant. If it isn't in contact with the propellant, it won't ignite the motor. It's held in place with a tiny wad of flameproof recovery wadding stuffed into the nozzle. Or it can be taped in place.

Don't "constipate" the motor by putting too much tape over the nozzle. Some model rocket motors require a clear nozzle when they ignite. Or, at most, they'll tolerate the minor pressure buildup caused by a wad of flameproof tissue stuffed in the nozzle.

Don't worry about the wire or wadding. Once the motor starts, the internal pressure of the motor quickly builds up to more than 100 pounds per square inch. Both the wire and the wadding will come off the nozzle in a hurry. In fact, a nichrome wire is good for only one ignition.

The nichrome wire must be of very small diameter in order to possess high electrical resistance. The smaller the wire, the higher the resistance. The higher the resistance, the less electrical current is required to get it hot.

Many beginners start thinking about capacitor–discharge ignition systems. Forget about it because here's how and why an electrical igniter or any hot wire electrical device works.

The amount of electrical current that will flow through any conductor is determined by Ohm's Law which is written:

$$I = E/R$$

where E = voltage in volts, R = the electrical resistance in ohms, and I = the current in amperes that will flow. The higher the voltage, the more current will flow. The higher the resistance, the less current will flow.

But it takes electrical *power* to cause an igniter to heat up. In an electrical circuit, the power is determined by the equation:

$$P = I^2 R$$

where I = the current in amperes, R = the resistance in ohms, and P = the power in watts. This equation tells us that if you increase the resistance, the power that's dissipated in the circuit to get the wire hot is halved. But, if you double the current in amperes, the power required goes up four times!

Together, these two equations tell us several things. First, if the voltage is too low, not enough current will pass through the igniter to generate enough power to raise the wire temperature to the point where the solid propellant will ignite. However, if you double the voltage, twice the amount of current will flow, and this will increase the power and the heat by a factor of four.

The electrical resistance of 2 inches of #32 nichrome wire is about 1 ohm. If you use a 6-volt battery and a good wiring system that itself has a resistance of 1 ohm, 16 watts of power will get to the igniter. This is just about the minimum required to get the igniter hot. However, if you use a 12-volt battery, it will have to deliver only one fourth the amount of current to achieve the same igniter temperature.

I made some tests of igniters using nichrome wire 0.010 inch in diameter (AWG #30 wire) and hooking some precision electrical measuring instruments into the circuit. I found that 5 volts is just barely enough to get a 2-inch nichrome wire igniter warm. 6 volts did a better job because the wire began to glow a dull red. At 12 volts, the wire glowed a bright red. At 18 volts, the nichrome wire burned in two before it got bright red-hot all over.

Never plug an igniter into a 120-volt AC house circuit. This is a direct short circuit. A nichrome wire igniter under those circumstances will allow some 90 amperes of current to gush through it. If it doesn't cause the house circuit breaker to open, it'll certainly wreak havoc with your electrical ignition system! And the nichrome wire will simply vaporize without getting hot long enough to ignite a model rocket motor.

Both Estes Industries, Inc. and Centuri Engineering Company have developed a special model rocket motor igniter that uses very little current and operates on 6 volts. Estes calls it the "Solar Igniter®" and Centuri's is the "Sure-Shot II®." Other manufacturers have also come out with low-current igniters. They require small, inexpensive dry cell batteries instead of the heavy lead–acid motorcycle or automobile batteries that model rocketeers used to have to lug to the flying field. It took twenty years, but we've finally divorced the model rocket from the automobile battery. I highly recommend these "Solar Igniters," "Sure-Shot IIs," and other types. Used properly and in accordance with instructions, they are efficient, reliable, and eliminate the need for big launching batteries.

However, an electrical ignition system and battery intended for use with these sensitive low-current igniters won't work with other kinds of igniters. The heavy-current igniters that have been used for years in model rocketry will ruin instantly the small dry cell batteries of these modern low-current ignition systems. These new systems simply won't deliver the amount of current required. But a system designed to "sock it to" an igniter will work perfectly well with the new low-current igniters. Therefore, the all-around model rocketeer, able to handle any motor and igniter, equips himself with a heavy-duty electrical ignition system.

There's one thing to remember when putting together an electrical ignition system, whether from a kit or from scratch: Do a good job, build it correctly, and don't take shortcuts. Your electrical ignition system will last for thousands of flights if you build it properly. But it will get tossed around in your range box, dropped in the mud, pulled through the dirt, stepped on, tripped over, pulled, yanked, and otherwise abused, deliberately or accidentally. Together with your launch pad, it takes more beating than any other system in model rocketry. And the electrical ignition system is your insurance of safe, successful ignition when and where you want it. It deserves lots of TLC (Tender Loving Care) both in construction, use, maintenance, and storage.

A properly designed and constructed electrical ignition system will last for years. Mine is twenty years old, and it still works. It must have very low electrical resistance so that it doesn't get hot instead of the igniter. It uses heavy wire, soldered connections, and heavy-duty components. But it's basically no different than the modern, low-current systems in the way it's wired and the basic components it uses.

The parts of an electrical ignition system are a launch controller, a battery or electrical power source, and connecting wires.

Basically, the launch controller is a switching device that makes sure no electric current gets from the battery to the igniter before you want it to do so. Then, when you complete its prelaunch program, it delivers enough current to the igniter to start the model rocket motor. Its basic part is a spring-loaded push-button switch that will remain in the "off" or "open" position and will automatically return to this condition when released. An ordinary electric light switch or knife switch will not make this automatic return to off, so these switches are extremely hazardous to use in your own home-built launch controller. Sooner or later you'll forget to return the switch to the safe off position after launching. The next time you put a model on the launching pad and hook up the igniter, the model will take off in your face. Having this happen to you only once is more than enough, believe me!

All commercial launch controllers have the proper spring-loaded, normally open, bush-button ignition switch. If you're building your own launch controller, you can use an ordinary doorbell button as shown in Figure 6-3.

Safety rules also require a means to disconnect the battery from the system or a safety key that performs the same function. The safety key in a launch controller physically opens the electrical circuit in addition to the ignition switch. It provides a double precaution, something engineers call "redundancy." When the safety key is out of the controller and in your hand or pocket, there's no way for any electricity to flow out to the igniter. Nobody is able to launch that model except *you*. Nobody can get cute and, as a practical joke, push the launch button just as you complete the igniter hookup. These are some of the reasons why you should have a safety key—although it isn't shown in Figure 6-3—and why you should *keep it with you*. Don't leave it in the launch controller. Don't keep it on a string tied to the launch controller. Keep it in your hand or in your pocket. Better yet, put it on a long red or orange ribbon around your neck or wrist. Or put a big, brightly colored ribbon or flag on it so you'll be able to see if you accidentally left it in the launch controller. The safety key is vital to your safety, and you shouldn't forget it for a single instant when you're on the flying field. With your safety key in your possession, you and *only you* can launch your model. You can work around the model on the launch pad knowing that nobody is going to play games with you and the launch controller while your fingers are under the model.

Once the safety key has been inserted into the launch controller, the controller is said to be "armed." This means it's ready to launch the model when you push the ignition switch.

A wiring diagram for a typical electric launch controller is shown in Figure 6-4. This is the basic circuit diagram for nearly all commercial electric launch controllers, too. You can see how the safety key opens the circuit. Notice that there is a "continuity light" wired across the ignition switch. This was invented by Vernon D. Estes in 1961 and is a small light bulb that lights up when you insert the safety key—if you've hooked up the igniter and everything is ready. The continuity light allows a very small amount of current to pass through the circuit, but not enough to get the igniter warm. It is a "limiting electrical resistance." When the ignition switch is pushed and closed, the light bulb is bypassed by the low-resistance circuit of the switch, allowing the full battery current to flow through to the igniter.

When using your launch controller with Estes Solar Igniters®, Centuri Sure-Shot II® igniters, and other types of low-current igniters, test before hooking up a model on the launch pad with one of those low-current igniters

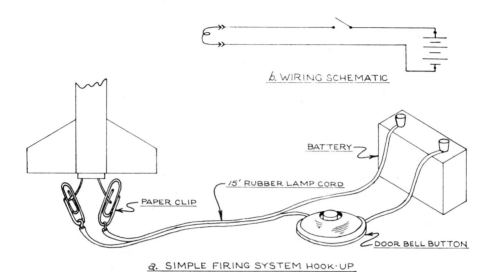

b. WIRING SCHEMATIC

BATTERY

/5' RUBBER LAMP CORD

PAPER CLIP

DOOR BELL BUTTON

a. SIMPLE FIRING SYSTEM HOOK-UP

Figure 6-3 Drawing showing the construction and hook-up of an inexpensive electrical ignition system.

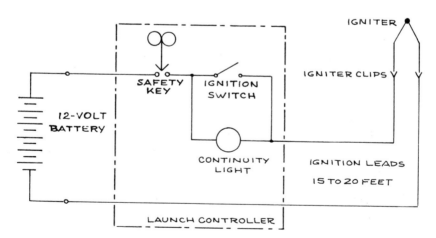

IGNITER

SAFETY KEY

IGNITION SWITCH

IGNITER CLIPS

12-VOLT BATTERY

CONTINUITY LIGHT

IGNITION LEADS
15 TO 20 FEET

LAUNCH CONTROLLER

Figure 6-4 The electrical wiring diagram of a typical electrical launch controller showing safety key and continuity check light. Nearly all commercial controllers are wired in this manner.

in it. Take one of the spare low-current igniters in your range box (You mean you didn't bring any extra igniters along just in case?) and hook the clips to it *without putting the igniter in a motor.* If the igniter doesn't activate when you put in the safety key, your launch controller design is safe to use with these igniters. If the igniter *does* activate, the easiest thing to do in the field is to remove the continuity light bulb. At your earliest opportunity, replace it with a type recommended by the manufacturer of the igniter. Write him a letter about it, and he'll be happy to tell you exactly what bulb to use; he may even have it available by mail order.

The continuity check light will *not* tell you that you don't have a short circuit between the clips or between the clips and any metal launch pad parts. You'll have to check that visually yourself when hooking up. The continuity light *will* tell you by failing to light up that you have dirty clips. a dead battery, or a number of problems covered in the troubleshooting section below.

Except for launch controllers designed for the low-current igniters, the wire used to connect the battery, controller, and clips—called the connecting wires or ignition leads—should be heavy-duty stuff. Don't use doorbell wire which is commonly available in most hardware stores. It's too small (AWG No. 22, usually) for use in heavy-duty model rocket electrical ignition systems, has far too high electrical resistance to high current, and is *solid* wire that will easily break after being flexed several times. And don't use "loudspeaker cord" which is AWG No. 22-2 insulated stranded pair available in hi-fi and radio shops. You want heavy-duty wire that will flex again and again. The best wire for this is also commonly available: rubber-covered or plastic-covered AWG No. 18-2 lamp cord, commonly called "zip cord." Or you can cut up a long extension cord made of the same double wire. Rubber-covered cord is preferable because the rubber insulation flexes easily over a wide temperature range; plastic-covered zip cord gets sort of stiff in the cold of wintertime flying.

You'll need something to hook the ignition leads to the igniter ends. Don't use big, heavy "alligator" clips because their weight may pull out the igniter. You can also use ordinary, garden-variety *paper clips* as shown in Figure 6-3; they work fine, but you've got to solder the ignition leads to them. Commerical electrical launch controller kits come with tiny spring clips called "micro clips." Many companies make these; ask for Mueller No. 34 "Micro-Gator" clips or Radio Shack 270-373 copper-plated micro clips. They're cheap, so get some spares because they'll eventually become cruddy, bent, and well-used after a couple hundred flights, and you can easily replace them, even on the flying field, *if* you have spares available.

When constructing your electrical launch control system, even if you're building a commercial kit, build it to last. Solder all connections, if possible. Don't just wrap wires together; solder them together or clamp them together

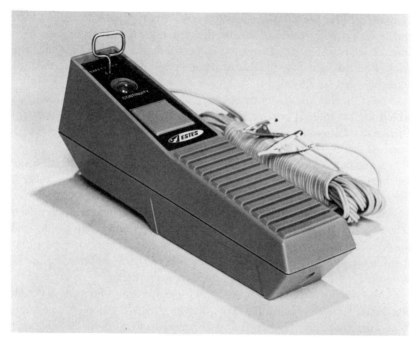

Figure 6-5 A typical electrical launch controller, the Estes Solar® system with internal Size AA penlight batteries for use with Estes Solar Igniters®. This unit can also be wired to use larger external batteries for nichrome hot-wire igniters. Note safety key and continuity check light. (Estes Industries, Inc.)

under a screw head. Remember, your system will have to pass enough electric current to operate a small electric oven or to operate a 500-watt light bulb. Build it so those little bitty electrons can get from the battery to the igniter with as little trouble as possible once you push the ignition switch.

Batteries still give beginning model rocketeers more trouble than anything else, even with the new low-current igniters. You can save yourself many of the headaches and heartaches of other model rocketeers if you'll take the time to read and heed what I have to say about batteries. I've rarely had a launching battery fail me; in those few instances when it did, it was my own fault.

When you push the ignition button on your launch controller, you are, in effect, putting a dead short across your battery. At the best, the resistance of the ignition circuit plus the igniter may be 2 ohms. To a 6-volt or 12-volt battery, this is indeed a short circuit. Ordinary flashlight batteries of the LeClanche zinc–carbon type were not designed for this kind of treatment and are totally ruined immediately.

Alkaline-type batteries will take it, however, Four Size AA alkaline dry cells are used for the Estes and Centuri low-current igniters. But don't try to use an ordinary hot-wire igniter with these specialized systems because those little dry cells just won't deliver the current to make a hot-wire igniter type work. Four to eight Size D alkaline batteries will do the trick, and they will sometimes last for fifty flights.

Next up the line in terms of size and power are the lantern batteries, which are an integral part of some commercial launch systems; they will operate low-current igniters for a long time and will otherwise last for several hundred flights with ordinary heavy-duty igniters.

Some model rocketeers buy the large "Hot Shot" type batteries designed for electric fences and such; they'll last for more than a year of regular weekend flying.

All of these batteries are LeClanche carbon–zinc dry cells or modifications of this type. Table 6 shows the most commonly available dry cells for occasional weekend flying. The number of flights you can expect from them depends on their age when you buy them, how long they sat on the store shelf, and the temperature on the day you're flying.

All batteries are chemical systems for converting chemical action to electricity. Like all chemical systems, they're dependent upon temperature. When they get cold, they work less vigorously and won't deliver as much current. A dry cell battery that works fine on a summer day may fail to ignite a model rocket motor in the dead of winter. Below 45°F, dry cells get cold, sluggish, and not very happy about supplying current for heavy-duty model rocket igniters. So keep batteries warm in the winter until you get ready to use them. Put them in a warm car until you've got the bird on the pad and are ready for the countdown. Another warming trick is to put the cold batteries inside your clothes and in your armpits to get them warm, although it's somewhat like putting ice cubes there. . . .

Among the finest model rocket ignition batteries are Size D sintered-plate nickle–cadmium rechargeable batteries (not pressed-plate ni–cads). They're very expensive, but they're lightweight and will throw a current of 50 amperes through a direct short circuit without the battery voltage collapsing.

Automobile batteries work fine for *all* sorts of model rocket ignition work. You don't have to remove the battery from the car, either; just connect the battery clips of your electric ignition system directly across the battery terminals in the car. Or you can buy an adapter that plugs into the cigarette lighter fixture. But the big problem with auto batteries is both their weight and the cost. Even if you buy a used battery from a service station (batteries that will no longer start a car will work an igniter in a dandy fashion), an auto battery is a mess to carry and store.

Motorcycle batteries are miniature auto batteries readily available at most

TABLE 6
Recommended Battery Chart

Type	Volts	Eveready	Bright Star	Burgess	Mallory	Marathon	Ray-O-Vac	RCA	Sears	Ward	Wizard
D Cell Energizer (8 required)	1.5	E95BP	7520	AL-2	Mn-1300	122	—	VS1336	4653	—	—
D Cell Photoflash (8 required)	1.5	850	10P	220	M-13P	124	210LP	VS736	—	3228	—
Radar-Lite	6	731	158	TW-1	M-918	896	918	VS317	4707	8MW	7D8918
Hot Shot	6	1461	146	S461	M-907	640	—	VS039	4668	7MW	7D8907
Lantern	7.5	715	155	4F5H	903	903	903	VS139	—	—	—
Lantern	9	716	164	4F6H	M-904	904	904	VS140	—	—	—
Radar-Lite	12	732	—	TW-2	—	732	926	VS342	—	—	—
Hot Shot	12	1463	187	2G8H	—	642	—	—	—	2335	—

Several manufacturers make the same type of battery as shown above. All those making a given type are listed.

cycle shops. They weigh about 5 pounds, are rechargeable, are easy to carry around, and work great.

An auto or motorcycle battery is technically known as a lead–acid battery. Properly cared for, it will last for years. You must keep it charged up because, sitting all by itself on the shelf, it will *self-discharge* over a period of several months. You must also keep the liquid electrolyte level at the specified point by adding only distilled water, never tap water. The demineralized water available in gallon jugs for steam irons works just as well and is cheaper.

When you buy a new lead–acid battery, it's "dry-charged," meaning that it doesn't have any electrolyte or battery acid in it. A container of acid is usually sold with a dry-charged battery. Battery electrolyte is a mixture of sulphuric acid and distilled water with a specific gravity of 1.280. It is put in the battery only once; thereafter, only distilled or demineralized water is added as needed to maintain the proper electrolyte level.

When you pour electrolyte into a new battery, *be careful*. The acid will make short work of your clothes if you spill any on them. If you get electrolyte on your skin, wash it off *immediately* with lots of water. Wash down any electrolyte spills with plenty of water.

All of this electrolyte mess and nonsense has been eliminated with the advent of the "gel cell" battery. This is a sealed lead–acid battery in which the electrolyte is in the form of a gel, which is a colloid like gelatin dessert. Other features built into the gel cell eliminate the need to vent it during charging—if the charging current is kept within limits, as discussed below—and the need to replenish the electrolyte. Gel cells are now widely used in other hobbies such as radio-controlled airplanes, cars, and boats. Being capable of high-current drain, they work fine as model rocket ignition sources. A 12-volt, 1.5-ampere–hour gel cell works beautifully for more than a hundred flights—although you should charge it up immediately after returning from each flight session. It's a plastic-cased battery 7 inches long, 1-3/8 inches wide, and 2-3/8 inches high weighing only 1 pound 10 ounces.

All lead–acid, ni–cad, and gel cell batteries are rechargeable, but the charging conditions for each are different. The normal charging rate for all these batteries is one-tenth of their ampere–hour rating. Thus, a 60-amp–hr car battery should be charged at 6 amperes; a 4.5-amp–hr motorcycle battery at 0.45 amperes; a 4.5-amp–hr ni–cad at 0.45 amperes, and a 1.5-amp–hr gel cell at about 0.15 amperes.

You can buy battery chargers of all sorts now. The high-current battery charger for an auto battery shouldn't be used to charge the sealed ni–cad or gel cell batteries because the high charging current may cause these batteries to explode. You'll probably be able to buy a battery charger specifically made to charge the type and size of rechargeable battery you have, and this is probably the best and least expensive way out. Otherwise, you'll have to make your own, and this can get somewhat complex.

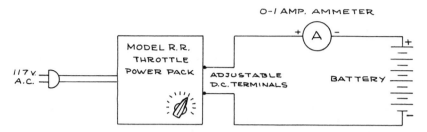

Figure 6-6 Wiring diagram for a simple battery charger using an HO train power pack for recharging lead-acid batteries and gel cells.

You may already have a battery charger that can handle motorcycle lead–acid batteries, ni–cads, and gel cells—and you probably don't know it. If you have an HO– or N–gauge model railroad set, you can use the power pack as a battery charger. This pack converts 120-volt AC house current to the 12- to 18-volts DC needed to run the trains. You'll only need an ammeter, an electrical meter for measuring current. Most "multimeters" (volt–ohm–millimeters) available from such places as Radio Shack can be set to measure the sorts of currents we're talking about for battery recharging. Radio Shack has such a multimeter available for less than $10. Or you can get a separate ammeter capable of measuring 0–1 amps DC (not AC) at hobby shops or radio parts stores.

If you use your model train power pack as a battery charger, wire it up with the ammeter as shown in Figure 6-6. *Watch polarity!* You can burn out your meter or ruin your battery if you don't charge with the correct polarity. Test for polarity this way: hook everything up, turn on the switch, and turn up the throttle control *just a little bit.* If the polarity is wrong, the ammeter will try to go off-scale *below* zero to the left. The remedy: Flip the train direction switch on the throttle pack or reverse the terminal connections to the battery.

When the polarity is known to be correct, turn the throttle control up until the ammeter reads the proper charging current for your battery. Check the reading every hour or so and adjust the throttle control to adjust the amperage. As the battery charges, the charging current will decrease, and you'll have to increase the throttle setting to maintain a constant charging current.

Charge lead–acid auto and motorcycle batteries at a constant rate of one-tenth of the capacity. Thus, a 50-amp–hr auto battery should be charged for 16 to 18 hours at *no more* than 5 amperes. A 5-amp–hr motorcycle cell should be charged for 16 to 18 hours at no more than 0.5 amperes. Fast charges build up heat in the battery which will warp the plates and dislodge active plate materials which, in turn, will fall to the bottom of the battery and eventually short out the plates. The state of charge of a lead–acid battery can be determined by measuring the specific gravity of the electrolyte in each cell. You can buy an inexpensive battery hygrometer at an auto supply store to do this. When all cells show a specific gravity reading of 1.280, the battery is charged. All lead–acid batteries produce gaseous hydrogen during the charging process, so charge them with the filler caps removed from the cells and with the battery in a well-ventilated place such as an open garage. Keep the electrolyte level in each cell at the proper point by adding only distilled or demineralized water. A lead–acid battery left to itself will self-discharge after several months of inactivity. A lead–acid battery that has become completely discharged or that will not take a charge has become "sulphated;" its plates have become coated with a hard, insulating coating of lead sulphate. A sulphated lead–acid cell can usually be rescued by slow-charging it for up to several weeks at a rate 1/100th of the amp–hr rating. I've saved many a "dead" battery this way.

Nickel–cadmium batteries are sealed cells, and there is no way to check the state of charge of these. The best thing to do is to keep Ni–Cd batteries on a constant "trickle charge" of 1/100th of their amp–hr rating at all times when you're not using them; this will keep them up to full charge all the time.

Gel cells must be charged differently, according to experts from the Academy of Model Aeronautics, NAR's sister organization for model airplanes. Limit the charging current to 1/5th of the amp–hr capacity until the charge voltage reaches 2.4 volts per cell (14.4 volts for a 12-volt battery). Then maintain a constant voltage until the charging current drops to 1/10th of the amp–hr capacity. Then disconnect because the charge is complete. Thus, for a 12-volt, 1.5-amp–hr gel cell, charge at 0.3 amperes until the charging voltage reaches 14.4 volts. Then keep reducing the charging current, holding a constant 14.4 volts charging voltage, until the charging current drops to 0.15 amperes. Once a gel cell is charged, it will hold its charge for several months, unlike an ordinary lead–acid battery.

7

Launchers and Launching Techniques

Good electric ignition is only one part of getting a model rocket off the ground for a successful flight. You must also have a launcher or launch pad.

A model rocket launcher is a device that holds the model rocket to be held in a prelaunch position so the igniter can be hooked up. It also restrains the model's freedom of motion once ignited until it achieves enough air speed to be stable.

If you were to set most model rockets directly on their fins on the ground with their noses pointed straight up and attempted to launch them in such a manner, they probably wouldn't fly. (There are a *few* exceptions, but don't go looking for them.) Sitting on the ground, a model rocket has no air flowing over its fins to permit the fins to stabilize the model and keep it going in the direction it's pointed. About a quarter of a second is required for most model rocket motors to build up thrust and accelerate a model rocket to a speed of about 30 mph where the fins can properly act to stabilize the model.

Without a launcher, a model rocket will probably topple over during the first split second of flight. When this happens, it goes all over the place. And you can't outrun it.

For safety, a launcher must *always* be used!

A rod launcher is one of the oldest, simplest, and most commonly used launchers. Various styles are available at reasonable prices from model rocket manufacturers. Some of them have electric ignition systems built right into them. And most of them have the following basic features:

- A launch rod is nothing more than a piece of 1/8-inch diameter hard steel wire at least 36 inches long. It's held in a vertical or near-vertical position by the launch pad base. The 1/8-inch by 36-inch rod is the NAR standard launch rod. In European countries, it's 3 millimeters in diameter by 1000 millimeters (1 meter) long.

  Most launch pad kits come with a two-piece rod. This allows the long rod to be put into a shorter package. It's also a lot easier to carry around and store. But with a two-piece rod, you're always in danger of losing or forgetting half of it. Half a launch rod will keep you from launching your models because you should *never* attempt to launch a model rocket with less than a 36-inch rod length.

If you want to make a two-piece rod into a one-piece rod, stick a 1/8-inch length of solder into the hole of one of the rods, push the other half of the rod into the hole on top of the solder, and hold the assembly over a match or candle (being careful to keep your hands well away from the hot area) until the solder melts to hold the rods together. Or you can go down to the hobby store and buy a 36-inch length of 1/8-inch diameter "music wire" which is very strong steel wire used to make the landing gear of some model airplanes. The longer steel launch rod is clumsier to carry around, but you're always certain of having a full-length rod with you.

Launch rods get dirty and rusty. Put a wad of steel wool in your range box and use it to polish the rod before each flight session. Or you can pay a visit to your local steel supply warehouse (look in the telephone Yellow Pages) to buy a rod made of stainless steel that will never rust; the one I got in 1967 is still the one I use, and I never have to put steel wool to it.

As a matter of fact, if you want launch rods of greater diameter and length for launching larger model rockets, you can buy these at most local steel supply warehouses, too. I have a 5-foot length of 3/16-inch diameter rod and a 6-foot length of 1/4-inch diameter rod . . . with modifications to my launch pads to accept these larger diameter launch rods for larger, heavier models.

• A simple but effective launch pad base is nothing more than a piece of 3/4-inch plywood about 12 inches square. Drill a 1/8-inch diameter hole vertically into the middle of the board. Tap the launch rod down into the hole.

Most sophisticated launch pads, including some now available from manufacturers, are made with a central piece and three legs that form a tripod. One or more of the legs are usually adjustable to permit the launch rod to be tilted or to compensate for uneven ground. A good launch pad base will have its legs spread widely apart for stability; when the launcher sits on the ground with a large model on it, it should be very difficult to tip it over. This is important when flying on uneven ground or on gusty days.

A launch pad should have a tilting mechanism so that the launch rod can be tipped slightly away from the vertical position. This allows you to put in a little bit of tilt to compensate for the effects of wind, primarily weathercocking of the model when it takes off in a wind. We'll discuss this in detail later. The NAR Safety Code says that you must always launch with the rod pointed within 30 degrees of the vertical, and all commercial launch pads are designed so that this limit cannot easily be exceeded. Tilting is accomplished by a movable joint at the base of the launch rod,

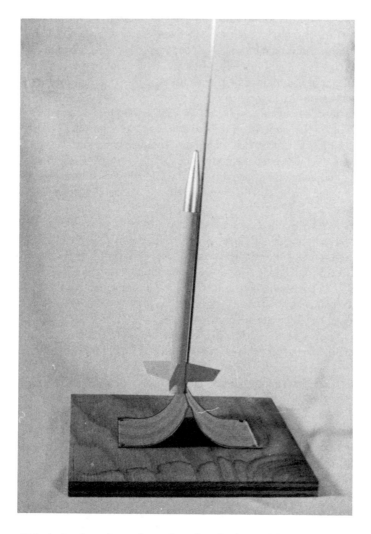

Figure 7-1 A simple rod-type launch pad with plywood base, launch rod, and jet deflector cut from a tin can.

by adjusting the angle of one or more of the launch pad legs, or by a tilt-leg adjustment on one leg. Always make sure the tilting mechanism is securely tightened so the model cannot tip over the launch rod with its own wieght. Also make sure the launch pad cannot fall over with a model in place and ready for launch. If necessary on windy days or with heavy or long models, put one or two rocks or bricks on the base or the legs to hold the launcher in place. Some modelers make U-shaped staples of

1/16-inch wire or from a coat hanger. They hold launch pad legs down by inserting the inverted-U over a leg and into the soil.

Some commercial launch pads are available with the ignition battery or batteries designed into the launch pad base. This makes the launch pad very stable and difficult to tip over because of the weight of the batteries placed low on the launch pad.

- A launch pad should always have a jet deflector to prevent the motor exhaust from striking the launcher or the ground. A jet deflector must be made from steel. In most kits, it's a steel stamping. Although a flat metal plate can and is used as a jet deflector, it can turn the exhaust jet back on the model in the split second before the model lifts off; this can discolor or burn the tail of the model or the fins, especially swept-back fins that extend below the body tube and motor.

A better jet deflector is a bent or angled piece of sheet steel that turns the jet at right angles and directs it away from the model and the launch pad. This is important if your flying field is covered with grass because such a deflector can go a long way toward preventing a grass fire being started by pieces of hot wadding and igniter debris that comes out of the motor immediately after ignition.

A jet deflector can be cut from a tin can. Always use a steel can, not an aluminum one. Deflectors made from aluminum will not stand up to the temperature of the jet exhaust. A hole will be burned through an aluminum deflector after only a few flights. A steel deflector will last for a long time—how long, we don't know, because I'm still using a steel deflector on a launcher I built in 1962.

- Another launcher feature that's very helpful is an umbilical mast or tower. This is a rod or dowel standing a few inches to one side of the launch rod. It has a clip, clothespin, or piece of tape for holding and supporting the weight of the ignition leads and clips. Some model rocket igniters have a tendency to be pulled out of the nozzle by the weight of the leads and clips when the igniter begins to get hot. Many, many misfires have occurred unnecessarily when the weight of the leads and clips pulled the igniter out of the motor before ignition could occur. An umbilical mast prevents these problems by supporting the major weight of the leads and clips, leaving only a few inches of firing leads for the igniter to support.

If I'm flying from someone else's launch pad that doesn't have an umbilical mast, I usually tape the ignition leads to the bottom part of the launch rod *below* the model's launch lug, or to one of the launch pad legs.

An umbilical mast is very important when launching front-motored boost-gliders and rocket gliders, as we'll see later.

- To prevent accidental eye injury, you should also make a rod cap for your launcher. This can be nothing more than an old, expended motor casing painted a bright, fluorescent color with a streamer or ribbon glued to it.

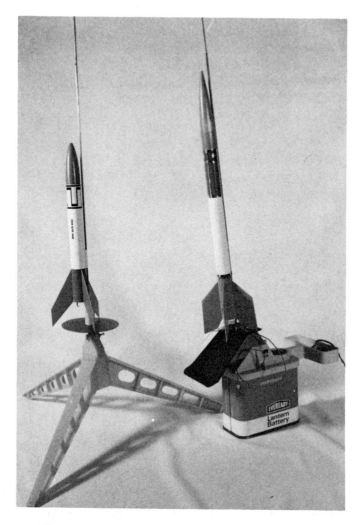

Figure 7-2 The commercially made launch pad at left is from Estes Industries, Inc. The commercial unit at right from Centuri Engineering Company includes battery and electrical system as part of the unit.

The rod cap is placed over the top of the rod when the launcher's not being used on the flying field or when it's being carried or stored. The rod cap alerts you to the location of the end of the rod so you don't run into it. You can also make a rod cap by attaching a streamer to a spring clothespin and clamping it to the top of the launch rod when the launcher's not in

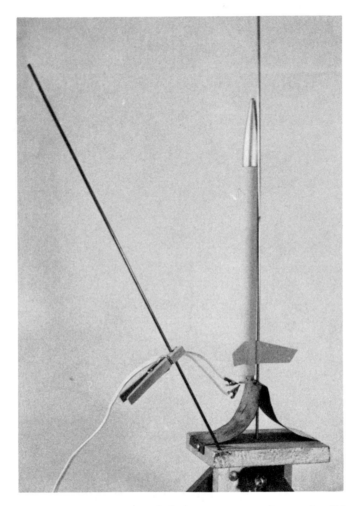

Figure 7-3 An umbilical mast and clothespin support the weight of the electrical leads and prevent igniter pull-out.

use. You can further protect your eyes from the end of the launch rod and from other things that might happen on the flying field if you'll wear sunglasses, which are helpful whether the weather be bright or cloudy. Dark green or gray lenses will help you see the model on a bright day, while yellow lenses are extremely helpful on a dull, cloudy day.

Be sure to remove the rod cap from the top of the rod before launching your model. One young rocketeer put his model on the rod and attached a clothespin to the top of the rod because he wasn't going to launch right

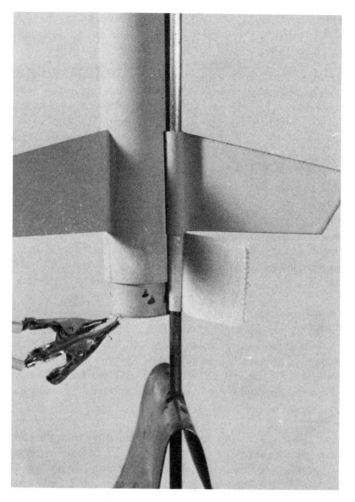

Figure 7-4 A model can be held up off the launch pad base by a piece of paper tape wrapped around the launch rod as shown.

away and he didn't want somebody to walk into the launcher. Later he forgot to remove it. The model stripped the rod cap off the end of the rod when it took off and then engaged in some rather spectacular aerobatics because the rod cap slowed the .model for a split second just as it was leaving the launch rod.

● If you fly from a grassy field, set your launch pad up on a tarpaulin or an old blanket. This will not only keep the knees of your pants dry and clean when you're hooking up your model, but it will also help prevent grass fires. The tarp or old blanket will catch any glowing pieces of wadding or igniter debris that might be ejected from the motor upon ignition. It's a precaution worth the trouble.

There are several helpful items you should have at the launch pad. Some of them can be tied to the pad with string so they don't get lost or "borrowed." One of these items is a roll of paper masking tape 3/4-inch wide for use when you need to tape things up on the pad. For example, some models may require support on the rod to keep their tails from resting on the launch pad base. Tape can be used as shown in Figure 7-4 to hold a model up on the rod. But don't forget to give the model all the rod length possible; it will be more stable in flight if you do.

At the launch pad you should also have a small square of No. 200 sandpaper or an emery board such as that used to file fingernails. This is known as a "JIC file." JIC stands for "Just In Case," and the file is used to remove the residue that forms on the igniter clips after several launchings. The file or sandpaper is used to clean this gunk off the inside jaws of the clips to expose a clean metal surface and make a good electrical contact with the motor igniter. In fact, in model rocketry parlance, you should "JIC the clips" before every launch.

I also keep a few spring clothespins in my range box or simply clip them to various parts of the launcher base where they'll be handy if I need them. They can be used in place of tape for supporting a model on the launch rod, for clamping ignition leads to the launch pad, and for handling the hundreds of other little clipping and clamping tasks that arise at the launch pad when you're hooking up models.

The simple, everyday things we model rocketeers use are often sources of amusement or consternation to nonmodelers. I'll not soon forget the time CBS sent a news team to cover a flying session of our model rocket club. The crew had just returned from Cape Canaveral and a major Apollo lunar mission. We put a 1/100th scale Saturn-V on the launch pad, and I went underneath it to hook it up. TV cameras and microphones caught every breathless prelaunch action and word. Tension among the TV people was as high as it had been at the Cape. I turned to a fellow model rocketeer and said in a matter-of-fact tone, "Please hand me that clothespin." The TV crew, used to the intensity, drama, and sophisticated high technology of a full-scale Saturn launch, came completely apart with guffaws of laughter while the director yelled, "Cut! Cut!"

The NAR Safety Code requires that model rockets be launched from launchers tilted no more than 30 degrees away from the vertical. Why this limit? One word that you've read before: *safety.*

It can be traced to an Italian Renaissance artillery officer and mathematician, Niccolo Tartaglia (1500?–1557) who discovered and wrote down what any artillery officer, naval gunnery officer, or ballistician still calls

TABLE 7

A. Altitude Correction Table

For computing change in air density as a function of elevation of launcher above sea level.

Elevation (in feet)	Multiply by
0	1.0000
1,000	.9710
2,000	.9428
3,000	.9151
4,000	.8881
5,000	.8616
6,000	.8358
7,000	.8106
8,000	.7859
9,000	.7619
10,000	.7384

B. Temperature Correction Table

For computing change in air density as a function of air temperature at launcher.

Temperature (in °F.)	Multiply by
30	1.0590
35	1.0486
40	1.0380
45	1.0277
50	1.0177
55	1.0078
60	.9980
65	.9885
70	.9792
75	.9700
80	.9610
85	.9522
90	.9435
95	.9350
100	.9266

Tartaglia's Laws of Gunnery. Tartaglia's Laws are as true for model rockets as for guns. Written in modern model rocket terms, they state:

1. A model rocket that will go 1,000 feet straight up will travel a horizontal distance of 2,000 feet if launched instead at an angle of 45 degrees.
2. When launched at an angle of 45 degrees, a model rocket that would go 1,000 feet vertically will ascent to a peak altitude of only 500 feet during its arcing flight.
3. The path followed by a model rocket when launched at *any* angle other than the vertical will describe a *parabola*. (For very long-range shots such as an ICBM or a space vehicle, the path becomes an *ellipse* because for long ranges the curvature of the earth must be accounted for. See any physics book or computer program for interplanetary trajectories of space vehicles.)

Tartaglia's Laws are correct for model rockets if you take into account air resistance. I tested them myself in the early days of model rocketry in the vast expanses of the American Southwest deserts under very careful safety controls (and before the NAR was even founded and the NAR Safety Code developed). This is one experiment you shouldn't perform. There are many other safer experiments, so launch within 30 degrees of the vertical as the NAR Safety Code specifies.

To answer a question nearly everyone asks at first: No, a model rocket will not fly horizontally. It has no wings to provide aerodynamic lifting forces against gravity, and its fins won't do the trick because they're too small and too far aft. When launched horizontally, a model rocket falls off the end of the horizontal launch rod, flops to the ground, and then skitters around on the ground during the thrust period of the motor. This performance usually removes the fins, bends or dings the body tube, and is not beneficial to a model rocket. It's also not very healthy for human bystanders because nobody has yet been able to out-run a model rocket.

The reason for permitting a launcher tilt of up to 30 degrees from the vertical is to allow the model rocketeer to compensate for the wind effects on his model during flight. In a wind, a model rocket will exhibit launcher tip-off just as the launch lug leaves the launch rod. The horizontally blowing wind makes the vertically ascending model rocket cock its nose *into* the wind like a weathervane or weathercock. And that's why we call the phenomenon "weathercocking." It happens with big NASA sounding rockets, too. The amount of weathercocking depends upon the velocity of the wind, the weight of the model rocket, the stability characteristics of the model rocket, the acceleration of the model, and other factors that would fill several pages with the complicated "banjo music" of partial differential equations.

But you don't have to know what a partial differential equation is to make weathercocking work for you. You can't fool Mother Nature, but you can usually make her work for you if you know how to do it. You can use

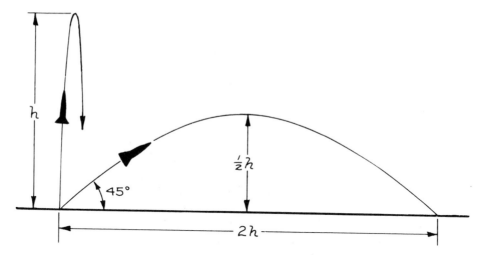

Figure 7-5 A graphic depiction of Tartaglia's Law, which is why launch pad tilt is limited to a maximum of 30 degrees from the vertical.

the weathercocking phenomenon to make your model rocket go where you want it to go and do what you want it to do.

To gain *maximum altitude* during a model rocket flight in windy conditions, tilt the launch rod *downwind* or *with the wind*. The amount of tilt is determined mostly by experience using the "wet finger" method of determining the wind direction and a knowledge of how your model performs in a wind. When the model lifts off, the weathercocking effect makes it tilt *upwind* or *into the wind*. As a result, the model ends up at a tilt downwind but quickly weathercocks into going straight up over the launch pad. It will deploy its recovery device at maximum possible altitude, so you'd better have lots and lots of open area devoid of trees and houses in the downwind direction.

For *maximum duration* flights or when flying from small fields, tilt the launch pad *into the wind* or *upwind*. This will make the model weathercock even more strongly into the wind. It then drives upwind so that the recovery device opens well over the upwind side of the field, the model drifts back over the field and launch site, and hopefully touches the ground just before reaching the rocket-eating trees on the downwind side of the launch field.

Most model rockets weighing less than 6 ounces (170 grams) can be launched safely from the standard 1/8-inch diameter, 36-inch launch rod. But when you start flying longer models such as the popular "superocs" that are "stretched" models ranging in length up to 6 feet, large payload-carrying models, models powered by Type E and Type F motors, or large-scale models, you should use 3/16-inch or 1/4-inch rods 5 to 6 feet long. Heavier models require more time and distance to build up flying speed, and you should give them all the launch rod possible for the best guidance. Long superoc models may have their launch lugs as far up the body tube as 24 to 30 inches, and this means that they've got less than a foot of standard length launch rod above the launch lug for initial guidance. There's absolutely nothing to be gained by using a short launch rod at any time, even though some ardent competition modelers claim that the friction of the launch lug on the launch rod decreases the maximum altitude of a model.

Naturally, when you use a larger launch rod, you have to put a larger launch lug on your model to fit over the rod properly.

Another kind of launcher seen on some model rocket ranges, especially during NAR competitions, is the tower launcher. This is a structure, often very simple in construction, in which a model slides between three or four vertical guide rails. The use of a tower eliminates the need for a launch lug on the model, and some experimentation has indicated that the aerodynamic drag of a launch lug can amount to 15 percent or more of the total drag on a model. A typical tower launcher is shown in Figure 7-7.

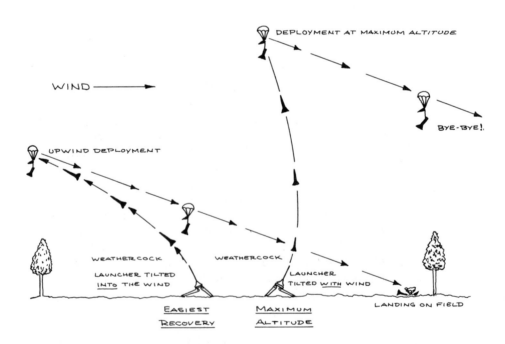

Figure 7-6 Launch pad tilt can be used to make the model fly where you want it to.

Figure 7-7 Tower launchers such as this one can be adjusted to accommodate body tubes of various sizes and eliminate the need for a launch lug. They are also stiffer than a standard rod for launching heavy models.

Figure 7-8 A C-rail launcher engages a standard launch lug on a standoff as shown and is stiffer than a regular rod.

There are many designs for launch towers, and they have both advantages and disadvantages. On the positive side, as was mentioned above, a tower eliminates the need for launch lugs, which reduces the total aerodynamic drag on a model. A tower also provides a very stiff launcher that doesn't sway or bend in the wind or whip as the model is launched. On the negative side, a tower is a complex piece of GSE that easily gets out of adjustment, especially when you carry it around in the trunk of a car. The guide rails must be carefully adjusted for each different model rocket body tube diameter. And some competition modelers actually believe (and some can produce data to prove their point) that the additional friction of three or four tower guide rails is greater than that of a launch lug sliding over a launch rod, a factor they claim cuts down on the maximum performance to an even greater extent than the additional drag of the launch lug.

(Model rocketry is a technology in miniature after all. In all fields of science and engineering, there is controversy. According to the famous aero-

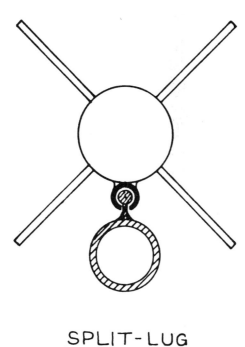

SPLIT-LUG

Figure 7-9 Cross section of a split-lug launcher and model.

dynamicist and rocket pioneer, Dr. Theodore von Karman, "How can we progress without controversy?")

Tower launchers are built from wood, plastic, or metal. The one shown in Figure 7-7 is typical of the launch tower types now in use, being nothing more than three 3/4-inch or 1-inch diameter wooden dowels 3 feet long attached in an adjustable fashion to a base. Some are designed to fit on inexpensive camera tripods so that the modeler can adjust the tilt as with a rod launcher.

A third kind of launcher is the rail launcher. It has also seen increasing use in model rocketry, especially with large and heavy models. In some cases, it has replaced the use of large launch rods entirely. It's often called a "C-rail" launcher because in cross section it looks like a squared-off letter C. See Figure 7-8. A model designed to use this sort of C-rail launcher can also be used on a standard rod because of its launching lug. For additional stiffness the C-rail can be screwed down to a long board or "strongback" which provides additional strength and resistance to whipping during launch.

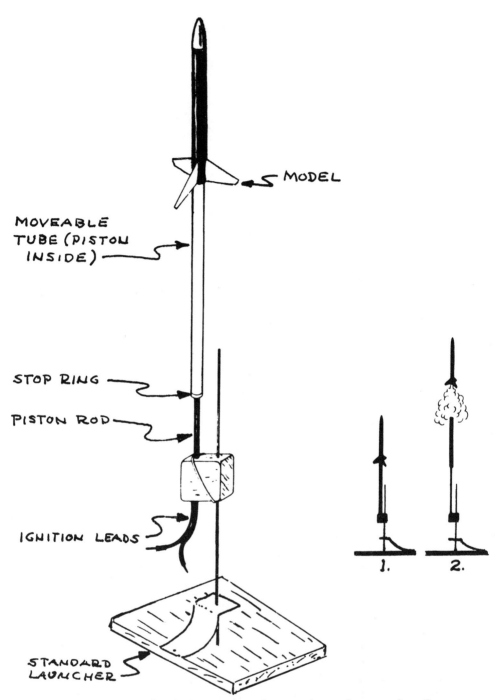

MODEL

MOVEABLE
TUBE (PISTON
INSIDE)

STOP RING

PISTON ROD

IGNITION LEADS

STANDARD
LAUNCHER

1.   2.

Figure 7-10 Howard Kuhn's piston launcher, an advanced concept from Competition Model Rockets, uses the exhaust gases from the model rocket motor to pressurize a tube-and-piston arrangement for rapid acceleration of a model at lift-off.

Another kind of launcher is the "split lug" type shown in Figure 7-9. The launch lug itself is split to clear the support for the attachments of the rod to the strongback. A model designed for a split-lug launcher can also be flown from a standard launch rod.

Since the early 1970s, competition model rocketeers have proven a launcher design known as a "closed breech" or piston launcher. It's still a tricky launcher to use and I don't recommend it for beginners at this stage in its development. Instead of allowing the exhaust from the model rocket motor to expand freely into the surrounding atmosphere, a closed breech or piston launcher uses the exhaust gas to pressurize an enclosed volume and move a piston to which the model is attached. Early piston launchers consisted of an enclosed tube into which the model was slipped atop a wooden piston that would come off when the model left the tube. Howard R. Kuhn invented the most popular form of the piston launcher which is widely used in competition today. In this piston launcher, the pressure tube (which is a standard body tube that slips over the lower 1/4 inch of the exposed motor in the tail of the model; the fit has to be *just right*) is attached to the model while the piston and its rod are attached to the launcher. When the motor is ignited (sometimes a tricky proposition to hook up and maintain hooked up inside that enclosed tube), the exhaust gas pressurizes the tube and forces the tube and model upward against the piston. At the limit of piston-tube travel— about 12 inches—the tube comes up against a stop ring, the model pops off the top of the tube, and the bird is launched. Early piston launchers had no guidance means for the model after it popped off the upper end of the tube; recent improvements to the design include 3-inch long wooden guideposts on the launcher tube that bear against the side of the model and provide some lateral restraint. The action of a pop launcher is spectacular because the model goes "pop" and is in the air almost immediately. The pressure of the trapped exhaust gases in the piston launcher results in very high initial acceleration.

At this writing, the only available piston launcher kit is made by Competition Model Rockets (CMR), headed up by Howard R. Kuhn himself. I strongly recommend that you buy, build, and use a CMR piston launcher to find out how it's built and how best to use it safely. It isn't an expensive unit, but the technology of piston launchers is an advanced area of model rocketry. As I mentioned earlier in this book, model rocketry has progressed so rapidly that it's impossible to go into all the details of many of the current advancements. If you're interested in piston launchers, get the CMR catalog, buy a CMR piston launcher from the man who invented it, and learn how to use it. Don't start out on your own. Why reinvent the wheel when you can learn how others have done it successfully?

8

How High Will It Go?

Once a model rocket leaves the launcher, it's a free body in space, even though it's still surrounded by the earth's atmosphere. It's been projected beyond the earth's surface, and its actions as a free body in space cannot be duplicated easily while it's on the ground. But we can account for the effects of the earth's atmosphere, and if we subtract these effects, we can study the motion of the airborne model just as if it were in outer space. We can discover where it will go, how far it will go, and how fast it will go.

It's possible to "fly a model rocket on paper." All you need are the elementary tools of simple arithmetic—addition, subtraction, multiplication, and division—and pencil and paper. Better have an eraser, too, because even the experts make mistakes. With these, you can find out in advance of flying your model exactly how it will perform. You don't have to be a genius to do this. And it's very exciting to work out all the numbers and then have the model perform the way the numbers said it would.

When the first edition of this book was published in 1965, hand calculators and home computers were nonexistent. Electronic digital computers were so big and so expensive that nobody but large corporations and universities could afford them. But since the fourth edition came out in 1976, home computers and hand calculators that can be programmed have become part of the equipment used in schools, business, and the home. These computers have revolutionized all of the formerly tedious calculations involved with studying model rocket performance, designing model rockets, and model rocket altitude tracking. Therefore, although we'll proceed through the basics of doing it by pencil-and-paper methods, we'll also discuss some simple computer programs written in the BASIC language that you can run on any computer capable of handling BASIC—or on a programmable hand calculator if it has enough memory storage and if you can revise the program to be compatible with one it understands. The reason for discussing the "old fashioned" pencil-and-paper methods is for the benefit of those who don't have access to computers yet. In addition, by looking at these "old fashioned" precomputer methods, you'll get an understanding of what goes on during a model rocket flight and, based upon that, be able to work any bugs out of your own computer programs.

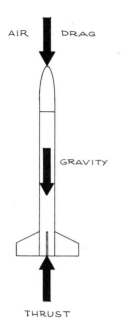

AIR   DRAG

GRAVITY

THRUST

Figure 8-1 Diagrammatic drawing
showing the forces on a model
rocket in vertical flight.

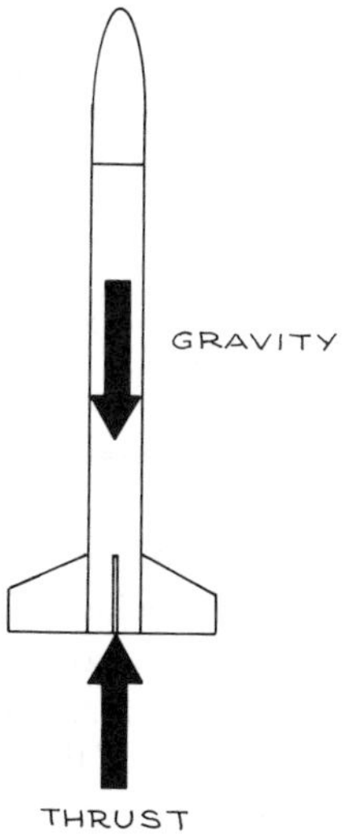

GRAVITY

THRUST

Figure 8-2 Diagrammatic drawing
showing the forces on a model rocket
in drag-free vertical flight.

(One of the disadvantages of the digital computer revolution has been
the growing inability of people to recognize whether or not the computer
answers are "in the ball park," to use the engineering vernacular. Does the
answer *look* reasonable? Is the decimal point in the proper place? These are
questions that immediately come to mind among those who were forced to
use the "ancient" methods of pencil-and-paper calculations or work with that
strange analog computer known as a "slide rule." By starting out with the
basics, you'll be able to determine "ball park numbers" in advance and make
a pretty good SWAG [scientific wildly assumed guess] at whether or not your
answers are basically correct and "close enough for government work," as the
saying goes.)

Although a model rocket has three flight phases—powered flight, coasting flight, and recovery—we're going to discuss only the first two phases here.
Recovery will be treated separately in another chapter.

Three basic forces act upon a model rocket in flight. You can think of
a force as the application of energy to the model in such a way that the flight
path is changed. As shown in Figure 8-2, and assuming *vertical* flight, these
three forces are:

1. *Thrust* from the model rocket motor that acts on the back of the model and makes it accelerate, or change speed.
2. *Gravity* that is a force trying to slow the model in its vertical flight and that acts in an opposite direction to thrust.
3. *Aerodynamic drag* caused by the model having to move through the air; this drag force also acts to slow down the model.

During powered flight, all three forces act upon the model. But during coasting flight, thrust is zero. Therefore, only gravity and aerodynamic drag act on the model.

The effects of aerodynamic drag are complex, as we'll see in a moment. The effects of thrust and gravity are simpler to handle. So, to begin with, we're going to assume that there is no aerodynamic drag on the model. In other words, the model is assumed to be a *perfect* aerodynamic shape, a fiction which, if it were real, would eliminate a lot of scientific and technical headaches among professionals. So we're going to ignore the effects of the earth's atmosphere and *pretend* that our drag-free model is acted upon only by thrust and gravity, as shown in Figure 8-3.

By calculating the flight of the model as if it were flying in the vacuum of space, and then determining the flight with aerodynamic drag included, we'll see how extremely important proper streamlining and shaping is because the drag-free flight will be very different from the flight with drag included. You'll see what a tremendous amount of aerodynamic force is exerted upon a model rocket in flight. Most people think that the air is quite thin and that its effects are negligible; try telling that to somebody who's been through a hurricane! A model rocket flies faster than the winds of a hurricane.

Although model rocketry is conducted in the metric system, we'll use the more familiar English system here because many of you may not yet be familiar with metrics and probably don't feel "at home" with that system of measurement. Most Americans are still more comfortable with feet, inches, pounds, and ounces. And it's a simple matter to convert from metric to English and back again. Just refer to the conversion chart given in Table 3 in Chapter 3 (p. 36).

To better understand what happens to a model rocket in flight, let's briefly review some of the basics about the motion of bodies in space.

When a body moves from point A to point B, it covers the distance S between the two points. We'll consider only simple motion in a straight line in one dimension; add other dimensions later if you want to. Since the body cannot go from point A to point B in zero time, it takes a finite period of time T to cover the distance S.

If S = 1 foot and T = 1 second, the body is said to be moving with a *velocity* of 1 foot per second. Velocity is therefore defined as the distance traveled divided by the time required for travel. During the next interval of

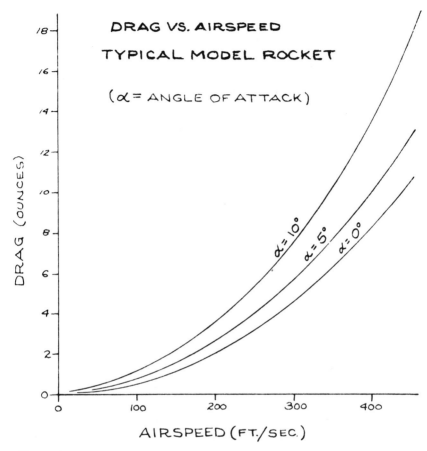

Figure 8-3 Drag force versus airspeed for a model rocket at various angles of attack. Data came from wind tunnel tests at United States Air Force Academy.

time, T_2, a body at constant velocity will cover another distance segment equal to S.

If the velocity of the body is not constant but changing, the body is said to be *accelerating*. It changes velocity at a given rate of change. A car accelerates if it goes from 0 to 60 miles per hour in 10 seconds, for example. If the velocity of our hypothetical body at the beginning of our observational interval of time is 1 foot per second, and at the end of that interval of time it's 2 feet per second, it's changed its velocity by 1 foot per second during that period of time. If the time interval is 1 second, the body experiences an acceleration of 1 foot per second per second, or 1 ft/sec/sec, of 1 ft/sec^2.

Acceleration can change, too. But don't worry about that at this point.

In 1687, the theologian and astrologer Sir Isaac Newton published a document entitled *Philosophiae Naturalis Principia Mathematica*. In this classic document, written in Latin which was the universal language of science in those days (today the universal language of science is Bad English), Newton revealed his famous three Laws of Motion. Simply stated in words, they are:

1. Law of Inertia: A body at rest will remain at rest or a body in motion will remain in motion in a straight line with constant velocity unless acted upon by an external force.
2. Law of Acceleration: Change in a body's motion is proportional to the magnitude of any external force acting upon it and in the exact direction of the applied force.
3. Law of Reaction: Every acting force is always opposed by an equal and oppositely directed reacting force.

Model rockets—and everything else in the universe that moves at speeds well below the speed of light—obey all three of these Laws of Motion.

But they don't seem to make sense, do they? It doesn't seem rational to assert that if you push a body it will keep on going forever and ever at the same speed and in the same direction you pushed it. Our everyday experience tells us that the body slows down and stops. But the reason it slows down and stops is the application of an external force, usually friction, acting upon the body to change its velocity in exact compliance with the Law of Acceleration.

A model rocket is one of the few accessible objects that obeys *all three* of Newton's Laws of Motion in a straightforward and easily demonstrated way.

A model rocket's flight is primarily determined by the Second Law or Law of Acceleration which can be stated in the scientific shorthand of mathematics as:

$$F = ma \tag{1}$$

where F = the applied force (from the rocket motor and from gravity), m = the mass of the model rocket, and a = the resulting acceleration.

If the applied force is doubled (that is, if the thrust is doubled) and the mass is kept constant (that is, the weight of the model remains unchanged), the acceleration experienced by the model will be doubled. If the force remains the same and the mass (weight) is doubled, the acceleration will be reduced by one-half.

From the basic Newtonian Laws of Motion and the basic equations of velocity and acceleration that come from them, the following general equations define the relationships between motion, distance traveled, velocity, acceleration, and time. If you're interested in how they were derived, you'll find the derivations in any high school physics text. Otherwise, just use them as they're presented here to help you determine the flight characteristics of a model rocket. Actually, these equations are greatly simplified because few of you probably know about another discovery of Sir Isaac Newton: calculus.

$$s = vt \tag{2}$$
$$V_{av} = (v_2 + v_1) / 2 \tag{3}$$
$$a = (v_2 - v_1) / t \tag{4}$$
$$s = v_1 t + (at^2) / 2 \tag{5}$$
$$2as = v_2^2 - v_1^2 \tag{6}$$

where s = distance, v = velocity, v_1 = velocity at the start of the time period, v_2 = velocity at the end of the time period, V_{av} = the average velocity during the time period, a = acceleration, and t = length of the time period.

Appendix II at the back of the book shows how you can use these equations to compute acceleration, velocity, and distance traveled by a model rocket.

These equations and the method in Appendix II are also very easy to program in the BASIC language of computers, but this is a simple problem for a computer. We'll present a BASIC computer program that takes into account *all* of the factors involved in model rocket flight, including changing thrust, weight, and air drag; here, a computer is *really* helpful.

A model rocketeer is primarily interested in only a few things about his model rocket's flight, such as maximum altitude and coasting time. These permit him to determine the field size he'll need and the motor type he should use for the flight. Maximum altitude and coasting time can be computed for the drag-free condition by a very simple method if you have three pieces of information about your model rocket:

1. You must know the *total impulse* of the model rocket motor.
2. You must know the *burnout weight* of the model—the takeoff weight minus the propellant weight.
3. You must know the *duration* of the model rocket motor thrust.

You can get the total impulse from the type code of the model rocket motor or from the manufacturer's specifications in the instruction sheet or his catalog.

The weight of your model rocket can be determined by weighing it on a postage scale or, more accurately, on a laboratory balance in the school science lab. Weigh it with a loaded motor and wadding installed, the exact condition it will be in at lift-off. The burnout weight can then be determined by subtracting the propellant weight—you can find that data in the instruction sheet or the manufacturer's catalog—from the maximum or gross lift-off weight you've just measured.

The duration of the model rocket motor thrust is obtained from the motor manufacturer's specifications given in the instructions accompanying the motor or in his catalog.

Let's run through an example here. Suppose your model rocket has the following characteristics:

$$\text{Total Impulse of motor } (I) = 2.25 \text{ newton-seconds}$$
$$0.506 \text{ pound-seconds}$$
$$\text{Duration of motor } (t_b) = 0.32 \text{ seconds}$$
$$\text{Lift-off weight of model } (W_0) = 30.1 \text{ grams}$$
$$= 1.06 \text{ ounces}$$
$$= 0.066 \text{ pounds}$$
$$\text{Propellant weight } (W_p) = 3.12 \text{ grams}$$
$$= 0.11 \text{ ounces}$$
$$= 0.0069$$
$$\text{Burnout weight of model } (W_b) = W_0 - W_p$$
$$= 30.1 - 3.12 = 26.98 \text{ grams}$$
$$= 0.95 \text{ ounces}$$
$$= 0.0594 \text{ pounds}$$

Note that we've started with all units in the metric system and have converted them into the English system. All units are in terms of pounds and seconds in the English system, including force. (One of the problems in dealing in the English system is that there is no separate unit for force as there is in the metric system.)

Recall that by definition the total impulse is the total change in the *momentum* of a body. Momentum is mass times velocity. Written as an equation, it is:

$$I_t = m_1 v_1 - m_0 v_0 \tag{7}$$

where I_t = total impulse of motor, m_1 = mass at burnout, v_1 = velocity at burnout, m_0 = mass at lift-off, and v_0 = velocity at lift-off.

Since at zero time, or at the instant of launching, the model rocket's velocity is zero—it's sitting at rest on the launch pad—the equation reduces to:

$$I_t = m_1 v_1 \tag{8}$$

Now when you weighed your model rocket, you actually determined the *force* with which it was being pulled toward the center of the earth by gravity. To determine its *mass*, you must divide its *weight* by the acceleration of gravity (32.2 feet per second per second) to get its *mass* which is in terms of *slugs*:

$$m = W/g \tag{9}$$

So Equation (8) becomes:

$$I_t = (W_1 v_1)/g \tag{10}$$

Transposing to get v_1 over to the left side of the equation by itself, we get:

$$v_1 = (I_t g)/W_1 \qquad (11)$$

The term v_1 is equal to the velocity at the end of the impulse or thrust period. Therefore, it's the burnout velocity of the model rocket. It's also the maximum velocity the model can attain during its powered flight and its coasting flight. Therefore, we also call it V_{max}. And for our hypothetical experimental model rocket we can calculate:

$$
\begin{aligned}
V_{max} &= (0.506 \times 32.2) / 0.0594 \\
&= 16.29 / 0.0594 \\
&= 274.3 \text{ feet per second}
\end{aligned}
$$

This is the maximum velocity attained by the model rocket. It's equal to 187 miles per hour.

Remember that I said model rockets were the world's fastest models? And this example was propelled by a low-powered Type A motor!

We must now compute how high the model is at burnout (S_b). We can use Equation (3) and Equation (2) given earlier.

$$
\begin{aligned}
V_{av} &= (v_1 + v_2) / 2 \\
&= (0 + 274.3) / 2 \\
&= 137.15 \text{ feet per second}
\end{aligned}
$$

$$
\begin{aligned}
S_b = vt &= V_{av} t_b \\
&= 137.15 \times 0.32 \\
&= 43.88 \text{ feet}
\end{aligned}
$$

Now you know why recovery devices are not deployed at burnout of the model rocket motor! At burnout, our hypothetical model is only about 44 feet in the air and is traveling at 187 mph. It hasn't achieved it's maximum altitude yet, and it's moving at a speed that would tear its recovery device to shreds.

When the motor stops thrusting and the time delay starts to work, the model enters the coasting phase of its flight. It coasts upward to maximum or peak altitude (apogee), trading its momentum (velocity) for altitude. During coasting flight, only gravity and aerodynamic drag forces are acting on the model. Since we're ignoring aerodynamic drag in this example, only gravity is acting. The model is actually "falling upward" in a gravity field; it's in zero-g or weightlessness. It's being acted upon only by the acceleration of gravity, 32.2 feet per second per second. We can now go to Equation (6) and find out how far upward it will coast:

$$2as = v_2^2 - v_1^2 \qquad (6)$$

Since $v_1 = 0$, then $2as = v_2^2$, and:

$$s = V_2^2/(2a)$$

where s = altitude, v_2 = maximum velocity, and a = acceleration. This results in the coasting altitude (S_c) calculation:

$$S_c = V_{max}^2 / (2g)$$
$$= 274.3^2 / 64.4$$
$$= 75240.49 / 64.4$$
$$= 1,168.3 \text{ feet}$$

The maximum altitude (S_t) will then be the burnout altitude (S_b) plus the coasting altitude (S_c):

$$S_t = S_b + S_c$$
$$= 43.88 + 1,168.3$$
$$= 1,212.2 \text{ feet}$$

But what would happen if we put a Type B motor with twice the total impulse into the model and flew it again? The new model parameters would be:

$$\text{Total impulse} = 4.5 \text{ newton-seconds}$$
$$= 1.01 \text{ pound-seconds}$$
$$\text{Duration of motor} = 0.83 \text{ seconds}$$
$$\text{Lift-off weight} = 36 \text{ grams}$$
$$= 1.27 \text{ ounces}$$
$$= 0.079 \text{ pounds}$$
$$\text{Propellant weight} = 6.24 \text{ grams}$$
$$= 0.22 \text{ ounces}$$
$$= 0.0137 \text{ pounds}$$
$$\text{Burnout weight} = 26.76 \text{ grams}$$
$$= 1.05 \text{ ounces}$$
$$= 0.0656 \text{ pounds}$$

Running through the same calculations quickly, we find:

$$V_{max} = (1.01 \times 32.2) / 0.0656$$
$$=$$
$$V_{av} = 494.76 / 2$$
$$= 247.88 \text{ feet per second}$$
$$S_b = 247.88 \times 0.83$$
$$= 205.74 \text{ feet}$$
$$S_c = 495.76^2 / 64.4$$
$$= 245777.97 / 64.4$$
$$= 3,816.43 \text{ feet}$$
$$S_t = 205.74 + 3,816.43$$
$$= 4,022,17 \text{ feet}$$

Quite a difference! We *doubled* the total impulse, but the performance didn't simply double, did it? Although the model was slightly heavier at lift-off because of the additional propellant in the Type B motor, this heavier lift-off weight was more than offset by the motor's longer burning time. This pushed up the burnout altitude and also caused a greater burnout velocity. When we compare the two performances, the following points stand out:

1. If we *double* the total impulse, the maximum velocity increases almost *four times*. In other words, the maximum velocity increases as the *square* of the total impulse.
2. If we *double* the total impulse, the burnout altitude increases almost *four times*. Again, the burnout altitude increases as the *square* of the increase in total impulse.
3. If we *double* the total impulse, the maximum altitude increases about *four times*, too.

   In summary, performance increases roughly as the *square* of the increase in total impulse! This is an important relationship to remember.

   By running off another set of simple calculations for yourself, you can also see for yourself that any increase in weight of the model will decrease the altitude performance.

   Although the numbers we've just calculated work out nicely, they don't seem to jibe with the real universe, do they? From our own experience in flying small models of this sort with Type A and Type B motors, we know that the models simply don't go to altitudes over more than 1,000 feet with a Type A motor and more than 4,000 feet with a Type B motor. Generally, they go about 500 and 1,000 feet, respectively.

   What's wrong?

   We deliberately ignored the effects of aerodynamic drag in these simple flight calculations.

   And we can now sense the dramatic effect aerodynamic drag has on the performance of model rockets. Aerodynamic drag is a major force in model rocketry. And it's obvious that we've got to take aerodynamic drag into account when calculating flight performance.

   Believe it or not, air is considered to be a fluid. And the amount of air drag experienced by a model rocket can be calculated by an equation from the science of fluid dynamics, the basic drag equation:

$$D = 0.5\rho V^2 C_d A \qquad (12)$$

where D = drag force, ρ = air density, V = velocity of the model through the air, or the air past the model, C_d = a dimensionless number called the *drag coefficient*, and A = the frontal area of the model.

   The drag equation tells us:

1. Air drag increases as the air density increases. How can you change the air density to reduce drag? By going to a higher altitude where there is less air density. Or by flying on a hot day because air is less dense when it's hot.

2. Air drag increases as the *square* of the velocity. Double the velocity, and the drag force goes up *four times*. The faster the model goes, the greater the drag force becomes. See Figure 8-4.
3. Air drag increases directly as the drag coefficient increases. We'll discuss this point in detail because it's something you *can* work to reduce in model rocketry.
4. Air drag increases directly as the frontal area increases, and therefore as the *square* of the body tube diameter.

We can't do much about density except fly at high altitudes and/or on hot days. The amount by which air drag is changed by altitude and by temperature is shown in Tables 8A and 8B on page 285.

The value of the drag coefficient (C_d) depends upon many factors. Primarily, it depends upon the shape of the model. This includes the shape of the nose; whether or not there are any transition pieces or changes in body diameter; the number, shape, and airfoil of the fins; the location and size of the launch lug (if any); and the smoothness of the surface finish on the model. (The size of the model is taken into account in the frontal area term in the equation.)

The drag coefficient isn't constant. It changes with change in the model's angle of attack, the angle between the long axis of the model and the direction of the air flowing past. See Figure 8-5. For most model rocket shapes, the drag coefficient increases with increasing angle of attack as shown in Figure 8-6, which was derived from tests of a model rocket made in the wind tunnels of the United States Air Force Academy in 1958 (data never grows old). As you can see, the frontal area of the model presented to the oncoming airflow also increases with increasing angle of attack, and this increases the value of A in the drag equation, too.

What does all this mean to you as a model rocketeer?

If a model wobbles in flight, thereby flying at different angles of attack, its drag force will be greater than that of a model that slips through the air smoothly with little or no wobble. This is only one reason why you should build a stable model rocket.

There are methods that can be used by a model rocketeer to alter the drag coefficient; we'll go over these in detail in the following chapter.

But how do we work the drag force equations into the flight calculations?

Answer: With great difficulty or by the use of a small computer.

Before the small general purpose microcomputer capable of being programmed in BASIC came along, the calculations were long, tedious, and painstaking. This is because the calculations must proceed on a step-by-step sequential basis with the answers to one step providing the basic numbers to start calculating the next step.

Also, a lot of factors change during the flight of a model rocket.

Look at the drag equation again, and you'll immediately see that the

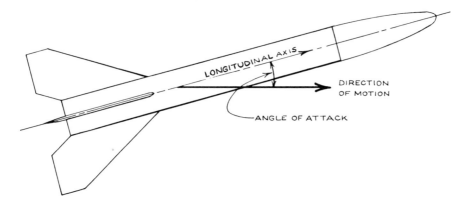

Figure 8-4 Graphic definition of angle of attack.

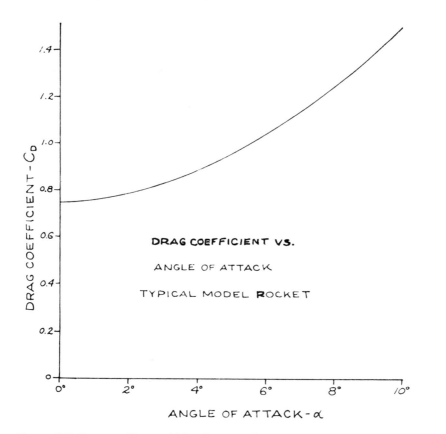

Figure 8-5 Drag coefficient (*Cd*) of a typical model rocket as a function of angle of attack.

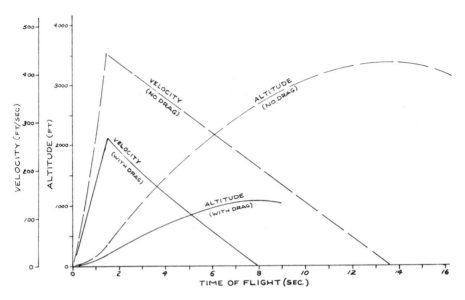

Figure 8-6 Comparison of computed drag-free flight performance with performance taking air drag into account.

drag force not only opposes the motion of the model through the air, but
also changes as the model's velocity changes. When the model takes off, its
velocity is low, and the model behaves as if there were no drag. However,
as the velocity begins to build up, so do the drag forces. This retards the
model even more.

Assuming a vertical flight, the acceleration of the model at any instant
during its powered or coasting flight can be calculated from the equation:

$$a = \frac{F - (0.5pV^2C_dA)}{W - W_0}g$$

Here, a = acceleration of the model during the time interval of the calculation
(usually made at intervals of 0.1 seconds); F = thrust of the motor at that
instant (it changes during operation, as you know, and becomes zero at and
following burnout); p = air density (which changes as the altitude increases,
but for the altitudes reached by model rockets this change is so slight that we
can consider air density to be constant); V = velocity of the model at the
start of the time interval; C_d = drag coefficient; A = frontal area of the
model; W = weight of the model at the start of the time interval; W_0 =
weight loss due to propellant burning during the time interval; and g =
acceleration of earth's gravity field.

I did this the long, hard way with pencil and paper *once*; that was enough. Then I got my computer, wrote a program in BASIC called "RASP-79E" which is reprinted in Appendix IV, and now let my computer do it in a matter of minutes. Go ahead and use RASP-79E, making whatever small changes in BASIC language are required by your particular computer. You can also use it for other model rocket motors by writing a separate but similar subroutine using data from the thrust–time curve of the motor you're interested in; often, manufacturers will change their motor performance characteristics—for example, since RASP-79E was written, Estes has substituted a B8- motor for the old B14- motor.

The use of modern digital minicomputers in model rocketry has permitted us to really get a solid handle on the effects of weight, size, drag coefficient, and motor type on altitude and time of flight. RASP-79E is accurate and has been checked in the field against flights of actual models tracked as described in a later chapter. In all cases, the field data matched the computed data within the limits of motor total impulse statistical variation (plus or minus 20 percent for Safety Certified motors, and plus or minus 10 percent for Contest Certified motors).

I ran the hypothetical model rocket of our example using RASP-79E in the same computer I used to write this book. I calculated for a Type A and a Type B motor with zero drag coefficient. The results were as follows:

Calculation	Motor	V_{max}	S_b	S_t
Book I_t	Type A	274.3 ft/sec	43.88 ft	1,212.2 ft
RASP–79E	Type A	234.03 ft/sec	51.16 ft	891.19 ft
Book I_t	Type B	495.76 ft/sec	205.74 ft	4,022.17 ft
RASP–79E	Type B	525.79 ft/sec	243.78 ft	4,516.65 ft

Why the discrepancies in the drag-free data? Our full total impulse method assumed *constant thrust* during the entire propulsion period; RASP-79E uses the thrust level of the motor as determined by static testing and computes the performance in intervals of 0.1 seconds. The actual thrust–time curve of the Type A8 motor is not steady, but more like the spike of a core-burner, and this additional real-world thrust level results in a higher burnout altitude but a lower maximum velocity. On the other hand, the Type B6 motor has a high initial spike but a long constant sustainer thrust; the combination of the spike plus sustainer gives us much higher drag-free computer calculations than the constant-thrust total impulse method.

However, the agreement is *close*, and that's what counts. Sometimes, if you can get your data to match within 10 percent, you can feel very happy about it.

The RASP-79E BASIC computer program can handle the change in thrust, change in weight, and change in velocity of the real world of model rocket flight. Because it computes the entire flight in time increments of 0.1 second, it's much closer to reality than programs that average various values. Using RASP-79E, I recalculated the flight of our hypothetical model rocket with a drag coefficient included. I chose a drag coefficient of 0.75 which seems to be about average for most sporting model rockets. The size of the body tube was established at 0.765 inches diameter. So our paper rocket is a small, inexpensive, single-staged sporting model similar to an Estes Mark III or Skyhook, or a Centuri Micron or Javelin. Let's compare the digested results of the RASP-69E drag-free ($C_d = 0$) and medium drag ($C_d = 0.75$) cases:

Motor	C_d	V_{max} (ft/sec)	S_b (ft)	S_t
A8-5	0	234.03	51.16	891.19
A8-5	0.75	227.32	50.43	457.41
B6-6	0	525.79	243.78	4,516,65
B6-6	0.75	460.78	226.93	1,120.88

These numbers are interesting and lead to some fascinating conclusions. First of all, air isn't as tenuous as we might think because it certainly seems to have a profound effect on the flight of something as small and streamlined as a model rocket. Secondly, the computer calculations with a reasonable drag coefficient included seem to be "in the ball park" because we know that a Type A motor will take a model to about 500 feet and a Type B will lift it up to a little over a thousand feet.

Of great interest as well is the fact that these modern computer analyses have caused us to change a few concepts we had about model rocket performance that were printed in every edition of this book up to the present one. Now it appears that we can draw the following conclusions about model rocket flight performance:

1. The aerodynamic drag of a model rocket has a very small effect on the burnout altitude of a model rocket unless high-powered motors are used which will in turn increase burnout velocity and thereby increase air drag and lower the drag-free burnout altitude. Aerodynamic drag *will*, however, greatly decrease the burnout altitude if the drag coefficient or frontal area are large.

2. Aerodynamic drag lowers the computed drag-free maximum altitude of a model rocket by 50 percent (for low-powered models) to as much as 80 percent (for high-powered models).

3. Aerodynamic drag forces on a model rocket become very large at velocities of 150 feet per second or greater, requiring very rugged construction for models designed for high performance and propulsion by Type D, Type E, and Type F motors.

4. The highest drag forces and the greatest structural stresses on a model rocket occur at or near burnout and rapidly become less as the model coasts up to apogee. In our hypothetical model, *negative accelerations* of 7.5 g's would be experienced due to air drag within a tenth of a second after burnout and rapidly drop to 2.6 g's a second later.

There is another interesting fact hidden in all these equations and calculations, and if you've got access to a computer you can use RASP-79E to discover it for yourself. The drag-free calculations would lead you to believe that the lighter the model rocket, the higher it will go. This just isn't true when you bring aerodynamic drag into the picture. In the drag equation, both the drag coefficient and the frontal area are divided by weight. Therefore, it's perfectly possible (and I've done it) to reach a point where an ultra-light model rocket acts just like a feather—and you can't throw a feather very far! The model's area-to-weight ratio or drag-to-weight ratio get to be so great that the aerodynamic forces completely overwhelm the momentum (mass times velocity) forces. The bird staggers up to 30 feet with a Type C motor in it and stops dead at burnout. This gave birth to the competition event known as "drag racing," and it's just what the name implies. The two-model heat goes to the bird that scores two out of three points: (1) getting off the pad first; (2) going to the *lowest* altitude of the pair; and (3) touching the ground *last*.

On the other hand, if you make a model rocket too heavy, the thrust from the motor will be unable to accelerate it to a high velocity, and the peak altitude will be low.

Therefore, a model rocket designer gives a little here, takes a little there, trades this for that, and tries to reach a happy compromise. For some reason, the early model rocket designs were right at or very near the optimum trade-off between weight and drag. This was a fortunate circumstance because it permitted the historic models to operate quite satisfactorily, giving model rocketry time to develop the nuts-and-bolts of its technology before having to get involved with technical trade-offs that we can now compute with modern digital minicomputers.

Analysis of flight dynamics and trajectories is a favorite subject for college undergraduates who've been model rocketeers and have taken their hobby to college with them. The Massachusetts Institute of Technology's Model Rocket Society has done a lot of work in this area using their big computers. Cadets at the United States Air Force Academy have utilized their wind tunnels for further drag studies. Other work has gone on at Ohio State University, the University of Maryland, and Kent State University, carried out by model rocketeers who've decided to become professionals.

Yet this is an area where junior high school students can work with minimal mathematical tools and abilities, and where students can learn how to program and use computers.

It's possible today to predict with exceedingly accurate results the altitudes, accelerations, and velocities of model rockets. These numbers are quite helpful in designing model rockets properly.

And it's very exciting to see the model rocket you've designed and "flown on paper" perform in the real world just as you calculated it would. It's an example of how engineers and scientists can accurately predict the future by knowing how the universe works and how to make it work for them . . . and for the rest of us, too.

9

Stability

In the preceding chapter, we've seen that the earth's atmosphere, the air, plays a major role in the flight of a model rocket. Aerodynamic drag greatly reduces performance. But, by utilizing careful design techniques based upon almost a hundred years of scientific research in aerodynamics, we can *use* the air to stabilize the model, to keep it going in the intended direction, and to make it fly predictably. We can also decrease the aerodynamic drag and increase the performance of a model rocket by understanding how the air flows around it and creates the drag force.

We must always keep in mind that a model rocket is a free body in space after it leaves the launch rod. It's not attached to the ground in flight, and the forces acting upon it in flight cannot be easily duplicated on the ground.

In discussing the performance of a model rocket in flight, we acted as observers on the ground, watching the model with reference to the ground. We were standing still and watching the model move. To better understand aerodynamic stability and aerodynamic shapes, we must change our point of view and travel with the model in its flight through the air. In other words, we must become theoretical passengers in order to see and understand better what is happening.

A model rocket in flight can move in *eight* different ways. This is technically known as "eight degrees of freedom." For simplicity and ease in considering the motions, we can reduce any motion of a model rocket in flight to a combination of one or more of the eight basic motions shown in Figure 9-2.

Thrust moves the model rocket forward. It comes from the model rocket motor.

Drag opposes the thrust force and attempts to slow the model.

(The *gravity* force can come from any direction, depending upon the altitude and flight direction of the model. We've already taken the gravity force into account in calculating the flight dynamics of the last chapter.)

Yaw is a swinging motion of the nose to left or right.

Pitch, an up or down motion of the nose, is similar to the yaw motions. Because a simple model rocket is the same shape in both the pitch and yaw aspects, the two motions are usually lumped together in the term *pitch motion.*

Figure 9-1 The size and shape of model rockets determine the air drag they will encounter in flight. Some shapes have less drag than others.

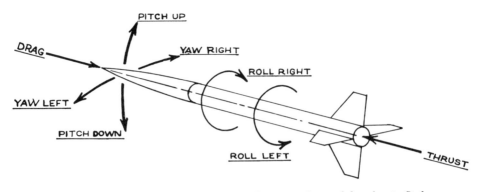

Figure 9-2 The eight degrees of freedom of motion of a model rocket in flight. Only two—thrust and drag—are linear; the other six are rotational.

But this isn't true for most boost-gliders and rocket gliders, as we shall see in a later chapter devoted to them.

Roll is a rotational motion where the model spins right or left about its long axis.

Thrust, drag, and gravity forces are *linear;* they produce motions in a straight line that are called *translational* motions.

Pitch, yaw, and roll are *rotational motions.*

Unlike a body experiencing a translational motion, anything that rotates must have a rotational axis, an imaginary line around which it spins. The earth, for example, has a rotational axis running through the North and South Poles. A model rocket has a roll axis that is an imaginary line running through the tip of the nose down along the centerline of the model and out the nozzle of the motor. There is also a pitch axis and a yaw axis, imaginary lines through some points in the model.

Where are these rotational points or centers of rotation? How do we find them?

When a free body in space rotates, it spins around an imaginary point where all its mass seems to be concentrated. This is its balance point. It is called the *center of gravity,* or CG.

You can perform a simple experiment to prove that a body rotates around a single point that's also its CG. Take a stick or wood dowel or body tube about 18 inches long. Balance it carefully and mark the balance point with

Figure 9-3 Rotation of a body about its balance point or center of gravity (CG).

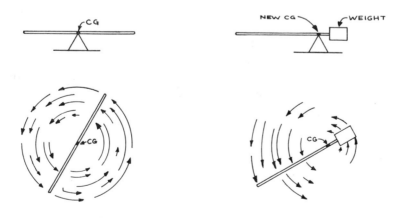

a felt-tip pen, making a line all around the stick or tube at that point. Toss the stick or tube into the air with an end-over-end motion. You will easily see that the stick or tube spins around the balance point you've marked. No matter how you throw the stick or tube to get an end-over-end motion, it will always rotate around this point.

Now put some putty or plasticine clay firmly on one end of the stick. Rebalance the stick and mark the new CG point, using a different color felt-tip pen. The CG will be in a new location, closer to the end on which you put the additional weight. Again throw the stick end-over-end, and you'll see that it now rotates around the new balance point or CG.

A model rocket in flight will rotate around its CG in the pitch, yaw, and roll axes. Why be concerned about these motions? Because if a model rocket rotates around its pitch or yaw axis, it's going to change its direction of flight. And you want that bird to go right where you pointed the launch rod: straight up. If it leaves the launch rod only to spin around its pitch and yaw axes, its angle of attack (and therefore its aerodynamic drag) will increase and the model will also go in a direction other than the way the launch rod is pointed. These other directions can often be very erratic and unpredictable, to say the least.

What can cause a model to rotate around its pitch–yaw axis? As Sir Isaac Newton said in a voice that calls down over the centuries to the spacemen of today: A *change of motion of a body can be produced only by an external force acting upon that body.*

There are many external forces that can cause rotational motions in a model rocket—gusts of wind blowing at the instant of launch, winds blowing at various altitudes during flight, fins crookedly positioned on the body tube, off-center thrust from the motor, irregularities in model construction, and many others. No matter how perfectly you build a model and no matter how carefully you try to launch it under ideal conditions, there will always be some tiny forces that will begin to produce pitch–yaw rotational forces the instant the model leaves the launch rod. You cannot eliminate them.

Therefore, the model must incorporate some sort of stabilization device that will overcome these forces, damp them out, and restore the model to its intended flight direction and altitude. Furthermore, this must be done very quickly—in a fraction of a second. A force must be almost instantly created to oppose the rotational force.

Most space rockets counteract rotational forces with an automatic electronic control system, an autopilot that uses gyros to sense the rotation in the pitch–yaw axes, and then sends electrical signals to a computer which in turn signals various hydraulic devices to tilt rocket motor nozzles or to fire attitude control rocket motors to correct the effects of the disturbing force. These control systems are, in comparison to model rocket technology, heavy, large, complicated, and expensive. Although some model rocketeers have experi-

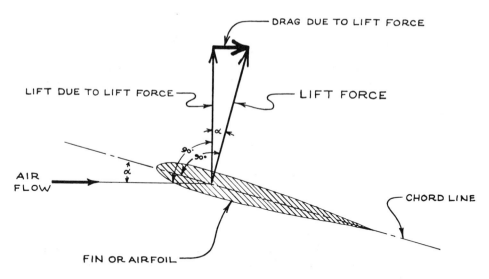

Figure 9-4 Lift force and drag force due to lift on a fin (Shown in cross section) at an angle of attack. Lift force is always at right angles to the chord line.

mented with tiny gyros for controlling their birds, most model rockets get their stabilizing and restoring forces in a simple manner from the air rushing past the model and acting upon aerodynamic surfaces—the fins.

When a moving stream of air strikes a surface broadside (at an angle of attack of 90 degrees) or even at a slight angle, it produces a high pressure on one side of the surface and a low pressure on the other, as shown in Figure 9-4. This pressure difference creates a drag force opposite to the motion of the air stream and a lift force that is at right angles (90 degrees) to the surface. The higher the angle of attack, the greater the lift and drag forces—up to the angle of attack where the surface "stalls." At the stall point, the air breaks away from the low-pressure surface, the lift forces decrease drastically, and the drag force increases tremendously.

By properly positioning the fins on the model rocket, by making them the right size, and by giving them the right shape, we can use this lift–drag force as a stabilizing force to offset pitch–yaw rotational disturbances.

As we saw with our tumbling stick, we can make the simplifying assumption that all the mass (weight) of a body is concentrated at its CG. It certainly acts this way, doesn't it? Let's extend this concept to *any* force acting upon the body, including the aerodynamic lift–drag force we've just introduced. Therefore, we can think of the body as having a point where all air pressure forces act. We call this point the *center of pressure*, or simply the CP.

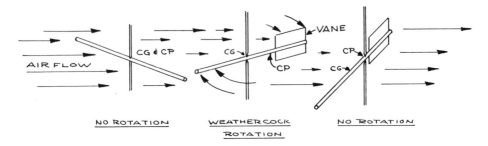

Figure 9-5 The stick-and-vane model under various conditions described in the text.

The function of the CP can be illustrated by another experiment with the stick or tube. If we grasp the stick at its balance point between a pair of pointed, low-friction pivots as shown in Figure 9-5, and hold it in the moving air from an air-conditioning duct (*not* from a fan because of the turbulence generated by a fan) or out of the window of a car traveling at about 20 miles per hour, we'll see that the stick doesn't rotate in the pivots due to any air pressure forces. No matter how you hold the stick in the airstream, it doesn't rotate. Obviously, there are no off-center or unbalanced forces to make it rotate, so the CG and the CP must be the same. We'd expect this because of the symmetrical shape of the stick.

Now glue or staple a piece of stiff cardboard to one end of the stick. Because of the additional weight of the cardboard vane, rebalance the assembly to find the new CG. Pick up the stick-plus-vane with the pivots at the new CG and place it in the moving air again.

The stick will immediately swing around with the vane downwind and the stick pointed directly into the wind. Congratulations! You've just reinvented the weathervane.

The presence of the cardboard vane produces more air pressure force on its end of the stick. This air pressure force was caused by the lift–drag of the vane.

The presence of the vane also put the CP at a different point than the CG. There's now a difference between the points at which two different forces act.

If you push the stick slightly to the side with your finger, you'll discover that the air pressure force becomes greater as you displace the stick more and more from the nose-into-wind position it naturally seeks to maintain. There is very little force when the angle of attack is low, but more force is created as the angle of attack increases. This is the sort of restoring force that will stabilize model rockets.

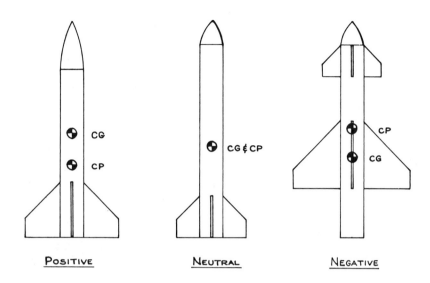

Figure 9-6 The three stability conditions with their CG–CP relationships.

If we start to move the pivot point closer to the vane end of the stick, we'll eventually find the point where the stick will no longer pivot into the wind. This is the point where the air pressure forces on both ends of the stick are equal. We've just located the CP of our stick-vane model.

We can replace the stick and vane with a model rocket, and it will behave in the same manner. When picked up and supported by pivots at its CG, the model will always point into the airstream.

With both models, the CP is *behind* the CG, making it an aerodynamically stable situation.

What would happen if we balanced either of our models so that the balance point, the CG, was behind the CP, and then picked it up with the pivots at the new rearward CG location? Very interesting! The model would try to fly tail-first. A real model rocket under thrust doesn't want to fly tail-first because of the thrust force. It becomes wildly unpredictable in flight, thrashing all over the sky because it's an unstable model.

As shown in Figure 9-6, there are basically three stability conditions for any body, including a model rocket—and only one of them is desirable for a model rocket. They are:

1. *Positive stability*, where in a model rocket the CG is ahead of the CP. It

Figure 9-7 The easiest way to determine the stability of a model rocket is to conduct a swing test.

has large fins set far back on the body tube. It will fly straight when launched and will weathercock into the wind at launch.

2. *Neutral stability*, where the CP and CG lie at the same location on the model. This might be caused by a lightweight nose or by fins that are too small, or both. There are no stabilizing and restoring forces present in the model during flight. It's free to wander anywhere in the sky, and some of its wanderings may be wild and certainly unpredictable. It may become stable or unstable at any moment because of the burnoff of the propellant, and then it might keep right on going in the direction it happens to be pointed at that instant.

3. *Negative stability*, where the CG lies behind the CP. In this case, the aerodynamic forces on the fins try to make the model fly tail-first, which it doesn't want to do under power. Once the nose swings in pitch or yaw after leaving the launch rod, a force exists to keep it swinging. The unstable model usually pinwheels end-over-end and winds up going nowhere except to flop to the ground.

Remember the positive stability condition by the mnemonic or memory aid: C before P. This is the alphabetical stability rule, because the CG comes before the CP in a stable, flyable model rocket.

Figure 9-8 Determining the CP of a model using a cardboard cutout of a side view of the model, thus determining the center of lateral area.

You should *never* fly a model rocket other than kit designs until you have determined its CG–CP relationship to ensure that it will be stable in flight.

The easiest way to make this determination is by the swing test. Tie a 4-foot length of string around the body tube of the model at the balance point. The model with a loaded motor installed should hang in a horizontal attitude from the string. In other words, the string is around the model at its CG. Hold the string in place with a bit of tape so the string doesn't slip.

Then, making sure that nobody's in the way, start to swing the model around your head in a horizontal circle. The longer the string, the more valid this test. If the model is stable, its nose will point in the direction you're swinging it. If the nose points elsewhere, you must add some weight to the nose to bring the CG forward.

Don't be dismayed when I suddenly reveal to you here that the CP of a model rocket usually depends on its angle of attack. In other words, the model may exhibit good stability if it swings to a small angle of attack. But if it is displaced to a large angle of attack, or if a strong gust of wind hits it in real flight conditions, it may not be able to recover its stable condition. This is because the CP of most model rocket designs moves forward as the angle of attack increases. It reaches its most forward point when the angle of attack is 90 degrees. Of course, some model rocket designs have less CP movement than others.

How does a model rocket designer handle this mess? Quite easily.

In the early days of model rocketry we determined the CP location by letting it equal the center of lateral area. To determine this point, we made a cardboard cutout of the side-view shape of the model. When we balanced this cardboard cutout, it gave us the center of lateral area, which we used as the CP. Actually, the center of lateral area is indeed the CP of a model rocket if it's flying at an angle of attack of 90 degrees, the worst possible condition.

All of the models designed with this cardboard cutout method of locating CP flew successfully. But there were a few models that just "happened" and weren't really designed. They flew well even though the cardboard cutout method of CP location indicated they wouldn't be stable.

Model rocketeers knew from NASA technical reports that aeronautical engineers and fluid dynamicists had a method of computing CP, but it was very complicated and was used to determine the supersonic CP of rocket shapes, which was something of considerable concern in those days (and still is). Nobody seemed to be able to apply it to model rockets flying at subsonic speeds.

The breakthrough came in 1966 when James S. Barrowman, a professional aerodynamic engineer with the Sounding Rocket Branch of NASA's Goddard Space Flight Center who was a model rocketeer and also became president of the NAR later, presented a simplified technique for computing the actual CP of a subsonic model rocket at low angles of attack.

The Barrowman Method was extremely successful. I've even used it to calculate the CP of full-sized rocket vehicles for some studies on launching dynamics and how to insure that the rocket remained stable through the subsonic phase of flight.

The equations and procedure of the Barrowman Method is shown in Appendix III. These may seem to be "hairy" at first, but a lot of model rocketeers have mastered the Barrowman Method.

Thanks to modern digital minicomputers capable of being programmed in BASIC, you can do design work on very complicated model rockets. I've written a BASIC program, "STABCALC-1," for doing this. It will handle a model rocket with up to three transitions and up to three sets of fins. Thus, it will handle the stability calculations for three-stage models as well as simple ones. For simple models, the program doesn't bother to calculate transitions or fins you indicate aren't there. STABCALC-1 is presented in Appendix V; you may have to modify it slightly to conform to the conventions of the particular form of BASIC your computer uses, but these shouldn't be extensive. There's only one square root calculation involved, and the rest is simply arithmetic operation.

If you don't want to bother with any of this, you can get by just fine by using the good old cardboard cutout method. Your models will have too much fin area and will have a tendency to weathercock more in a wind, but they will fly in a stable manner if designed using the cardboard cutout method.

THE QUESTION OF STABILITY

How much stability should a model rocket have? How far behind the CG should the CP be located?

It's generally agreed among advanced model rocketeers and has been generally confirmed by flight tests that the CP should be no less than one body diameter behind the CG. In other words, if the body tube diameter is 1.34 inches, the CP should be a least 1.34 inches behind the CG. This is known as "one caliber stability." In gunnery terms "caliber" refers to body diameter, and the word comes to us from the days when rockets were part of the artillery corps of armies.

Technically, one-caliber stability is all you really need for most sporting models. It allows for the rearward movement of the CP with increasing angle of attack up to a reasonable limit beyond which your model probably won't go anyway. If your model has more than 2 to 3 calibers stability, it may be overstable and suffer from excessive weathercocking (which may be something you want in a parachute duration type of contest model). It will certainly have fins that are too big, adding additional unneeded weight.

What do you do if your model rocket checks out as neutrally stable or negatively stable—or if it is less than one-caliber stable? That depends on the model. It may be possible to increase the fin area or to move the fins back; this moves the CP rearward. Such an approach may not be possible with a scale model where the size and location of the fins are predetermined by the dimensions of the real, full-sized rocket. In the case where you can't change the size, shape, or location of the fins, you must add weight to the nose of the model to bring the CG forward. Sometimes you must do a little bit of both to obtain the optimum weight and the least amount of fin area. This is one of the things that makes a model rocket designer's work so fascinating. There are so many trade-offs that can be made.

For many years—and even now in some parts of the world where model rocketry, or "space modeling," is new—model rocketeers believed they should make the smallest, lightest model rockets possible in order to achieve maximum performance. As we've seen in the previous chapter, there's a serious flaw in this logic because it's possible to make a model rocket so lightweight that it tries to fly like a feather. This same make-it-smallest philosophy also resulted in some short, squat, fat little model rockets that were barely large

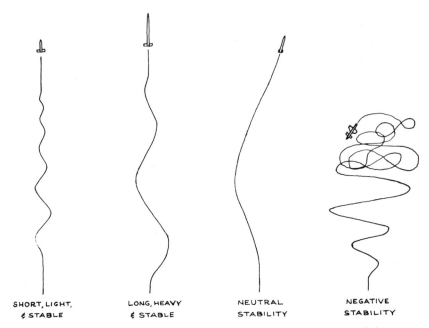

SHORT, LIGHT,   LONG, HEAVY   NEUTRAL   NEGATIVE
¢ STABLE   ¢ STABLE   STABILITY   STABILITY

Figure 9-9 Flight paths of model rockets with various static and dynamic stability characteristics.

enough to enclose a motor, a very small recovery streamer, and a nose. To obtain the proper stability, these models had fins that were sharply swept back behind the model. True, they had 1-caliber stability. But all of the major weights of the model—motor, nose, and recovery device—were located very close together.

In flight, these models wobbled back and forth very rapidly as they ascended. We found out earlier that drag increases as the angle of attack increases. Therefore, these squatty short models experienced substantial drag because of their constantly and rapidly changing angle of attack.

On the other hand, many of us had started to fly long, slender models that weighed more but that would slither upward with little or no wobble.

The basic performances are shown in Figure 9-9.

To understand why the two designs functioned as they did, let's go back to our stick model held between low-friction pivots out of a car window, our moving wind tunnel. Make a short stick about an inch in diameter and about 4 inches long. Put a cardboard vane 2 inches by 4 inches on it as shown in Figure 9-10. Make a longer stick about an inch in diameter and about 12 inches long. Cut a cardboard vane 1 inch by 2 inches and staple it to the stick as shown.

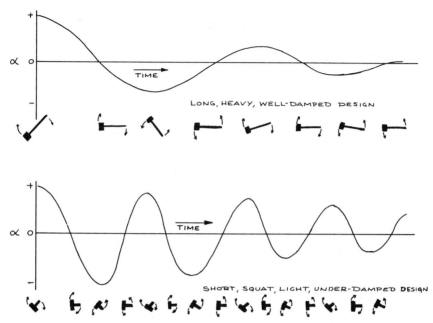

Figure 9-10 Visualization of dynamic stability with the stick-and-vane models described in the text.

Grasp the short stick model by the pivots at the CG and put it into the airstream. Displace it in the yaw direction with your finger and watch how it behaves. It's a stable model. The air hits the vane and produces a restoring force. The stick starts to swing back to zero angle of attack. It does so very quickly because all its mass is concentrated near its CG. Very little inertia is involved. The stick will reach its maximum rotational velocity when the angle of attack is zero, and it will not stop swinging. It will swing through zero angle of attack and take up an angle of attack on the other side of zero. The stabilizing situation is reversed, and the model then starts to swing back again. It will oscillate, swinging back and forth on either side of the zero angle of attack point. The oscillations will be very rapid and several will be required before they damp out and the model stops swinging.

Now hold the long stick with the small vane in the airstream in the same manner. It's also stable. Displace it in the yaw direction as you did with the short model. When released, the long model will also start swinging back, but it will take longer to do so. Its rotational velocity will be much less as it swings through zero angle of attack. It will also swing to the other side. But its oscillations through the zero point will be slow, will be fewer, and will damp out after perhaps only one or two passes through zero angle of attack.

These actions are caused by very complex phenomena grouped together under the general classification of *dynamic stability*. Just because we can hang a name on it doesn't mean that we understand it any better. Basically, the difference between static stability and dynamic stability is that of balancing a nonmoving device versus a moving one.

It is perfectly possible for a model rocket to have its CG and CP in the proper locations for good static stability while actually being dynamically unstable in flight. Model rockets encounter several kinds of dynamic stability problems:

1. A *statically stable but dynamically undamped model*. This is represented by the short, squat, fat little model that wobbles excessively as it flies.
2. A *statically stable but dynamically unstable model*. In this model, the CP and CG may be properly located with respect to one another, but the model is too heavy and has fins that are too small. These small fins, while creating a statically stable model with a CP behind the CG, are too small to produce enough restoring force to return the model quickly to zero angle of attack. They aren't large enough to overcome the turning momentum of the model. By the time these small fins can create sufficient restoring force, the model is doing something else.
3. A *statically stable but dynamically overdamped model*. In this condition, the model might have a long, skinny body with fins that are far too large. When the model weathercocks or rotates in the pitch-yaw axes, the fins produce too much restoring force and stabilize the model too quickly, causing it to fly as if it were almost neutrally stable. This condition is rarely seen, but it can exist with some canard-type boost-gliders where the wings are large and at the rear of the model.
4. A *model with pitch-roll coupling*. This is a weird form of instability that can really frustrate you if you don't know about it. In this situation, the model has some roll that's induced by fin, nose, or motor misalignments. At some point in the flight, the frequency of the roll becomes the same as the frequency of the motion back and forth in the pitch-yaw axes. The model will start to exhibit a "coning" motion where it's spinning about the roll axis while at the same time it begins to rotate in both pitch and yaw about the CG. The model can become completely unstable as it rolls madly around its long axis and spins horizontally end-over-end, going nowhere. The problem of pitch-roll coupling occurs in full-scale fin-stabilized rockets.

These dynamic stability problems were first mentioned in an earlier edition of this book. We really didn't know very much about them then. They were complex and were not yet actually understood by professional rocket engineers, either! There was a lot of discussion among rocketeers, professional and model, but nobody really had a handle on *any* of the dynamic stability problems. (If you think these are bad for simple rockets, they're life-

and-death matters for people who ride in asymmetrical airplanes that suffer dynamic stability problems involving such things as flutter and yaw damping!)

As a result, a model rocketeer named Gordon K. Mandell tackled the subject while he was still an undergraduate at the Massachusetts Institute of Technology. He improved the MIT low-speed, low-turbulence wind tunnel and developed some ingenious instruments to measure motion and forces on model rockets placed in that wind tunnel. He also made some simplifying mathematical assumptions and succeeded in linearizing the equations to the point where they could be useful in producing results for model rocket designers. His thesis on the subject was published in the very advanced book, *Topics in Advanced Model Rocketry* by Mandell, Caporaso, and Bengen, first published by MIT Press in 1973 and now reprinted by the NAR.

Mandell discovered that most of our model rocket designs were *overdesigned* on the safe side. This wasn't surprising since most of us tend to be a bit conservative in design because of all the unknown factors that we deal with and all the unforeseen conditions of a model rocket flight. No real rules of thumb have yet been developed from Mandell's classic work. Perhaps you or some other model rocketeer will pick up where Mandell left off so that these rules of thumb can be reported in a future edition of this book.

Briefly, Mandell stated some basic points of design to keep in mind:

1. Maintain a length-to-diameter ratio of 10 to 1 or more to provide adequate damping.
2. Maintain a static stability margin between 1 and 2 calibers to prevent overdamping, but don't go below 1 caliber.
3. Hold the roll rate of the model as low as possible to prevent pitch-roll coupling, and align the fins carefully in an attempt to get zero roll rate for best performance.
4. If you must increase the linear dimensions of the fins to get proper static stability, increase the *span* dimension (the dimension outward at right angles to the body tube) because this will increase the restoring force rather than the distance of the CP from the CG, which in turn improves the dynamic damping.

We now have static stability well in hand and understood, and Mandell has made significant progress toward solutions of the dynamic stability problems we run into from time to time. But there's still a lot of work to be done by model rocketeers who get interested in this area of the hobby and who want their work to be recognized.

This one reason why model rocketry should never be considered just kid stuff. I have used Barrowman's work and Mandell's findings in design work on real sounding rockets. Professional rocket engineers at NASA have used Mandell's work because it's just as applicable to full-sized rockets as to our models. It took a model rocketeer to get the complexities of dynamic stability simplified to a workable point.

10

Model Rocket
Aerodynamics

Thus far, we've been discussing some areas of aerodynamics without delving deeply into the subject itself. But, as we've discovered, aerodynamics plays a large role in model rocketry because model rockets are stabilized by aerodynamic methods and because air drag has a major effect upon model rocket flight.

Aerodynamics is a branch of the science of fluid dynamics, but it's concerned with only one fairly complex, composite gas made up of oxygen, nitrogen, carbon dioxide, and a few other elements. This special gas is the earth's planetary atmosphere. Aerodynamics may someday be extended to other planets that have atmospheres such as Mars, Venus, Jupiter, and Saturn. Aerodynamics is involved with the way air flows around various shapes and with the forces created in this process.

Many years ago, backyard inventors and "aeroplane" builders learned that some shapes have less drag than others, and that some shapes create more lifting force than others. There's a tremendous amount of information available on the subject of aerodynamics, some of it going back more than a century. For example, some of the data used by James S. Barrowman in developing his simplified CP calculations came from research done during World War I on biplanes such as the Sopwith Camel and Spad. Thus, our Space Age hobby of model rocketry has technical roots that go back to the days of the Red Baron and Eddie Rickenbacker. All of this data is readily available to model rocketeers, and many of them have used it.

Air can exert considerable force, as we've seen in the previous discussions of flight dynamics and stability. Uncontrolled, this aerodynamic force can tear a model to pieces. Properly used, it can stabilize a model rocket. Much of the force generated by air in motion depends upon the shape of the body around which the air flows. If you put your hand out the window of a car at 55 miles per hour (carefully, please), and if you open your palm broadside to the airflow (that is, at a 90 degree angle of attack), you'll feel a definite push against your palm. If you make your hand into a fist, this semispherical shape will have less aerodynamic drag, and you'll feel less push from the air.

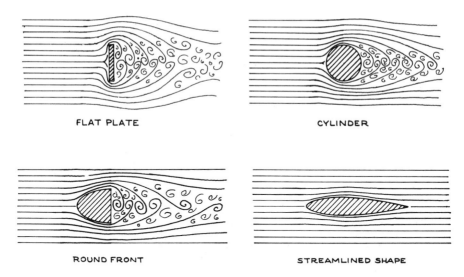

Figure 10-1 Typical airflow patterns around objects of various shapes.

The drag force is basically caused by the need for the object to push the air molecules out of the way, slide through them, and permit them to close in behind it again with the least amount of disturbance. Figure 10-1 shows how air can be visualized as flowing around some objects of different shapes. Thus, the shape of an object and the pattern of the airflow around it have a tremendous effect on the amount of drag force produced.

There are several forms of drag that are of interest to model rocketeers. They may be summarized as follows:

1. Friction drag.
2. Pressure drag.
3. Interference drag.
4. Parasite drag.
5. Induced drag.

Let's explore each of the above in turn, because drag is very important to model rocket flight.

Air is made up of molecules and is a mixture of gases. For our purposes in model rocketry, the molecules can be considered homogeneous air molecules rather than a mixture of molecules of different types. These air molecules are so small that trillions of them would fit on the period at the end of this sentence. You can think of the multitudes of molecules as tiny Ping-Pong balls separated from each other by very small distances. At the earth's surface under normal conditions of air pressure and temperature, the average

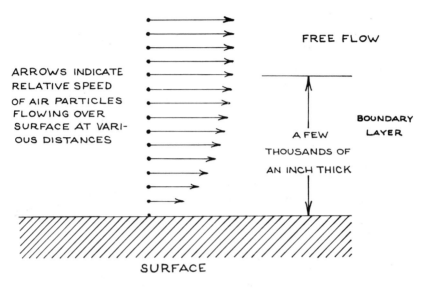

Figure 10-2 The boundary layer.

air molecule can travel only 0.000002419 foot (29 millionths of an inch) before hitting another air molecule. So things are crowded, and the Ping-Pong molecules are hitting one another all the time, creating a gross effect we call pressure.

When the molecules slide over the surface of a nose, body tube, or fin, friction results between the surface and the molecules. Actually, even though the surface seems to be smooth and shiny to the eye, it's full of microscopic hills and valleys. *Friction drag* is caused by the air molecules bumping into hills, rebounding off valley walls, and bumping into each other as a result. The rougher the surface, the more numerous are the microscopic hills and valleys for the air molecules to hit, and the greater the friction drag.

Actually, the airflow next to the surface exhibits some rather strange and unsuspected activity. Right on the surface, the friction and viscosity of the air slows the first layer of molecules almost to a standstill. The next layer slides and slips over the first layer at a little higher speed. So it goes, with each successive layer sliding faster over the layer below until the full free stream velocity is reached. This fluid flow phenomenon that takes place close to a surface is called the *boundary layer*. It's shown diagrammatically in Figure 10-2.

All objects moving with respect to the air possess a boundary layer. The thickness of the boundary layer varies with the size of the model and the speed of the airflow. If the layers are slipping over one another in an orderly fashion as shown in Figure 10-2, the average model rocket traveling at an air

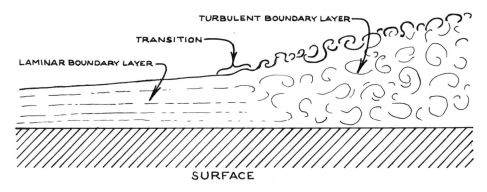

Figure 10-3 Boundary layer transition from laminar to turbulent state.

speed of 250 feet per second will have a boundary layer only about 1/1000 inch thick.

You can actually experience a boundary layer on a beach. If you lie down on the sand, you'll be in the boundary layer of the beach surface. You'll feel less wind speed there than when you're standing, and you'll be able to keep warmer on chilly days.

The boundary layer can be "laminar," as shown in Figure 10-2, or it can be "turbulent" as shown in Figure 10-3. When a boundary layer undergoes the transition from laminar to turbulent, or "transitions," the layers of air molecules in the boundary layer no longer slip easily over one another, but swirl and eddy about within the boundary layer itself. This makes the boundary layer become thicker, even though it's still attached to and flowing along the surface.

You can see laminar and turbulent boundary layers very easily by playing around with the faucet in the kitchen or bathroom sink, unless the faucet has an aerator. If it does, take the aerator off. Open the faucet slowly and carefully until the water streams out in a smooth, clear fashion. Then carefully open the faucet a little bit more. The stream will suddenly break into turbulence several inches below the spout. If you continue to open the faucet and increase the flow, the entire stream will become turbulent. Like air, water is a fluid and obeys all the same rules of fluid dynamics even though it has a higher density and will not decrease its volume under pressure.

With even the smoothest surface, a boundary layer becomes turbulent at some distance along a model rocket. This distance is a function of the size,

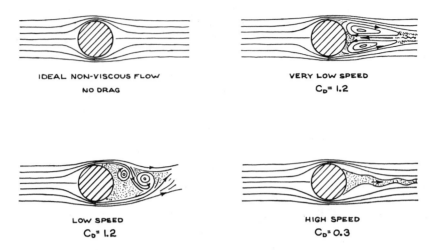

IDEAL NON-VISCOUS FLOW
NO DRAG

VERY LOW SPEED
$C_D = 1.2$

LOW SPEED
$C_D = 1.2$

HIGH SPEED
$C_D = 0.3$

Figure 10-4 Airflow around a cylinder under various conditions.

shape, finish, and speed of the model. Because of the swirling eddies of a turbulent boundary layer, friction drag is much higher than it is in a laminar boundary layer.

A very small irregularity on the surface will cause transition from laminar to turbulent boundary layer conditions. For most model rockets flying at 250 feet per second, a protuberance 0.0003 inch high will cause transition or "trip the boundary layer" from laminar to turbulent.

There is some data to indicate that the boundary layer on a model rocket usually trips on the nose, usually close to the nose base. Some modelers theorize that it's best to deliberately trip the boundary layer at the nose–body joint because it's going to trip there at the joint anyway. Therefore, they make the nose smooth but also give the body tube a glossy, mirror-like surface to reduce friction drag. There's some data to indicate they're right. Experiments conducted with a model rocket in a wind tunnel by Mark Mercer, a young rocketeer from Bethesda, Maryland, indicated a 24 percent increase in friction drag between a rough nose and a smooth nose.

To overcome the effects of friction drag, it's important that the entire surface of a model rocket be as smooth and glossy as possible.

The impact of tiny air molecules on the surface of an object, such as your hand sticking out of the window of the moving car, creates a drag force known as *pressure drag*. Obviously, this pressure on the front end of an object moving with respect to the air is caused by the impact of the air molecules on the object. But it's also possible to have *negative* pressure, a region of fewer air molecule impacts than the surrounding air. (Remember that the

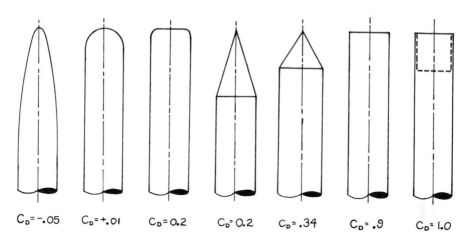

$C_D = -.05$ $C_D = +.01$ $C_D = 0.2$ $C_D = 0.2$ $C_D = .34$ $C_D = .9$ $C_D = 1.0$

Figure 10-5 Pressure drag coefficients for various nose shapes.

body is totally immersed in the earth's atmosphere, which produces a constant, static, ambient environmental pressure of about 14.7 pounds per square inch at sea level.) The region behind your hand in the auto airstream has negative pressure, a partial vacuum.

If there were no boundary layer, and if the air had no viscous characteristics that make it act like a very thin syrup, there would be no pressure drag whatsoever. This condition is shown in Figure 10-4. The air would move smoothly apart in front of the object, allow the body to pass through it, and then close in quietly and completely behind the object with nary a ripple. This is the case only for very small objects such as raindrops moving at very low speeds through the air.

An object the size of a model rocket moving through the air at 250 feet per second simply doesn't give the air enough *time* to slip out of the way and close in again behind it. In other words, the air must be shoved aside by the nose, an action that creates pressure drag and a boundary layer. If you've done a good job of design and construction, the boundary layer becomes turbulent at about the base of the nose and finally breaks completely away from the surface of the model at the blunt rear end where the air goes into swirling, eddying motion, completely unattached to any object. Thus a wake is created, just like that behind a ship in the water. The wake is the low-pressure area that tries to retard the motion of the model, and it must therefore be considered as a form of pressure drag. It's called *base drag* and is part of the total pressure drag of the model. For many model rocket designs, base drag is the major portion of the total pressure drag on the model.

It's possible to reduce the pressure drag on the front of a model to zero or negative pressure by shaping the nose properly. Many tests of "forebody shapes" were made in wind tunnels in the United States and Germany during World War II. The results of these tests are summarized in Figure 10-5. Just rounding the edge of a blunt nose by as little as 10 percent of the diameter of the nose itself can reduce the pressure drag enormously.

This wind tunnel data clearly tells us that the best low-drag nose shape for a model rocket isn't the sharply pointed "ogive" or conical shape of supersonic rockets, but the rather blunt-looking rounded-tip shape of the parabola of revolution or the easier made ellipse of revolution.

The rounded nose of a high-performance model rocket usually causes rookie model rocketeers to shake their heads in disbelief. "It doesn't look the way a rocket ought to look!" However, these doubters may not realize that the full-scale rockets are designed to fly with minimum drag at supersonic and hypersonic speeds where pointed noses are indeed low-drag shapes. Model rockets, on the other hand, spend most of their time at less than half the speed of sound (Mach 0.5). If you still don't believe that a rounded shape is best for subsonic flight, take a close look at the noses of subsonic jet airliners. The noses of Boeing, Douglas, and Lockheed jet airliners are rounded, and they cruise at about Mach 0.8.

Other wind tunnel tests have shown that the drag coefficient (C_d) of a parabolic or elliptical nose is a function of its length-to-diameter ratio as shown in Figure 10-6. The best length for a subsonic nose is between two and three times the base diameter. Extending the nose length beyond this doesn't appreciably lower the pressure drag and may actually increase the total drag due to the greater surface area on which friction drag can operate.

Usually, there's little that a model rocket designer can do to eliminate the base drag of a model rocket. Most models have a body diameter that's not very much larger than the model rocket motor casing itself. And the rear or nozzle end of a model rocket motor is very blunt indeed. Base drag is reduced during powered flight because the motor is thrusting, pumping billions of gas molecules into the base region. This is often evident after a flight by the slight staining of the base area by the yellowish-brown particles from the exhaust plume. However, the base drag is much higher during the coasting phase of flight.

To date, there is very little data to tell us just what the base drag really is during coasting flight. It's certainly not the pure base drag experienced by blunt-ended bodies in wind tunnels because the motor's time delay charge puts out gas which tends to relieve some of the low pressure area in the model's wake. We lack firm, experimentally verified numbers on this subject. It's a very good research project for a serious model rocketeer.

Some model rockets have body diameters that are greater than the diameter of the motor casing, perhaps because they have to carry large payloads

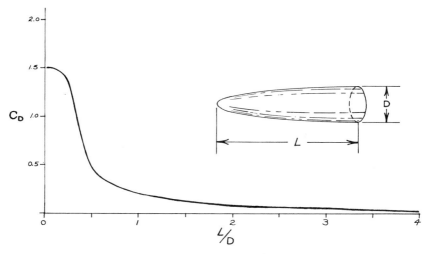

Figure 10-6 The pressure drag coefficient of a parabolic nose shape as a function of its length-to-diameter ratio.

or perhaps because the modeler just wanted to build a *big* model. The base drag of such large-diameter models can be very high.

The classic technical report on model rocket drag was written in 1967 by Dr. Gerald M. Gregorek of the Aeronautics and Astronautics Department of Ohio State University. Entitled "A Critical Examination of Model Rocket Drag for Use with Maximum Altitude Prediction Charts," Dr. Gregorek's report pointed out that the base drag of a large model rocket is equal to the cube of the ratio of the base drag of the model to the base drag of a model of the same shape but with a base diameter equal to that of the motor:

$$(\text{Diameter of the base/diameter of the motor})^3$$

George M. Pantalos, one of Dr. Gregorek's students, gave a dramatic example of this factor at work. If you have a model with a body diameter of 1.84 inches propelled by a motor with a diameter of 0.75 inch, the 1.84-inch diameter model will have a base drag 14.76 times greater than the same model reduced in size to the diameter of the motor.

To minimize the base drag of large model rockets, designers reduce the base area by tapering, or making a boat tail, as shown in Figure 10-7. According to Gregorek and Pantalos, the best taper angle is between 5 and 10 degrees with 6 degrees being optimum.

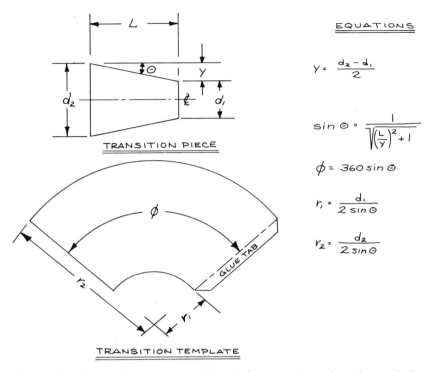

$$y = \frac{d_2 - d_1}{2}$$

$$\sin \Theta = \frac{1}{\sqrt{\left(\frac{L}{y}\right)^2 + 1}}$$

$$\phi = 360 \sin \Theta$$

$$r_1 = \frac{d_1}{2 \sin \Theta}$$

$$r_2 = \frac{d_2}{2 \sin \Theta}$$

TRANSITION PIECE

TRANSITION TEMPLATE

Figure 10-7 How to lay out a conical boattail or transition of any fore and aft diameter, angle, and length from a flat sheet of material.

Most model rocket builders make boat tails from stiff card stock paper, laying them out in accordance with the principles illustrated in Figure 10-7, which shows you how to make a tapered boat tail from a flat piece of paper. You can use this data to make a transition piece called a shoulder where the body diameter increases or a boat tail reduces the diameter of the model. The equations will permit you to make transition pieces with angles and lengths of your choice.

Additional work done by Pantalos in the Ohio State wind tunnel in 1973 showed there's an optimum place to put the boat tail or transition. Pantalos tested the three basic model shapes shown in Figure 10-8. Model A is a typical large-diameter model rocket with a constant body diameter. Model B is Model A with a boat tail at the rear of the body to reduce the diameter from that of the body to that of the motor. Model C has its diameter reduction or transition right behind that section of the model requiring the large-diameter tube for payload-carrying purposes.

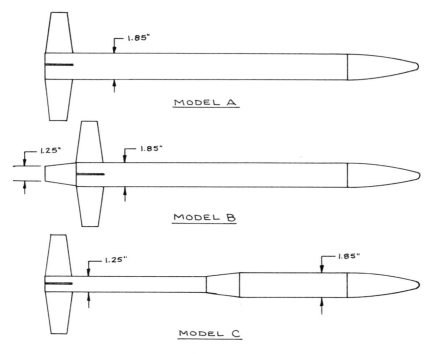

Figure 10-8 The three basic model rocket shapes tested for overall drag by Pantalos.

Which model had the least amount of drag when tested in the wind tunnel?

If you chose Model C, you're correct. It has the least body surface area and therefore the least amount of friction drag. Model B has the same base area and base drag as Model C, but greater surface area and therefore greater friction drag. Model A, of course, has more base drag than Model B and Model C, and more surface area than Model C.

See how these different forms of drag interact with one another and how designers must make technical trade-offs?

Interference drag is caused by the interruption of the boundary layer airflow over the body and the fins by the junctions between the body and the fins. That is, it's caused by the interference of the flow between two surfaces. Technically, this might be considered part of pressure drag, but we separate it from pressure drag in model rocketry because we face the question: Which model will have the least amount of interference drag? One with three fins in triform configuration, or one with four fins in cruciform configuration? The obvious answer is that the model with three fins will have 25 percent

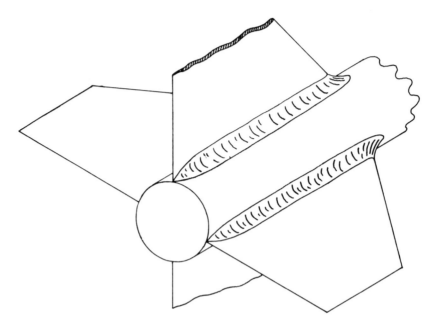

Figure 10-9 Fin fillets help reduce interference drag at the joints between fins and body tube. (Fillets are exaggerated for clarity.)

less interference drag than the model with four fins. This is why most high-performance model rockets have three fins.

This leads to an interesting story. The first three-finned rocket was the historic WAC-Corporal designed by the Jet Propulsion Laboratory of the California Institute of Technology in 1945. Before 1945, all rockets and bombs had four fins. Some aerodynamic experts didn't believe the three-finned WAC-Corporal rocket would be stable in flight. Dr. Frank Malina, the rocket's designer, quietly pointed out that arrows with three fletching feathers had been flying in a stable fashion for centuries. Often it pays to bridge the gulf that appears to exist between two apparently unrelated fields of human endeavor.

Interference drag can be reduced by the use of fillets, as shown diagrammatically in Figure 10-9. The optimum fillet radius should be 4 percent to 8 percent of the fin root chord, the dimension of the fin where it's glued to the body tube. A fillet that's too large will increase the surface area and hence the friction drag. A satisfactory fin fillet for most small model rockets can be made with a bead of glue.

Another method of reducing interference drag was suggested by Gregorek and confirmed by wind tunnel tests: Move the fins forward of the aft end of the body tube by a distance of about one body diameter (but be sure to check the CP calculation if you do). This slight forward location of the fins permits the airflow to smooth out along the body tube behind the fins before it enters the turbulent region of the model's base.

Parasite drag is drag caused by anything that sticks out from the body to interrupt the smooth flow of the boundary layer over the model. A major source of parasite drag on a model rocket (which is an exceptionally clean aerodynamic shape, by the way) is the launch lug that's glued to the body tube about halfway along the body tube. On an otherwise clean, streamlined, well-made model rocket, the launch lug's parasite drag can amount to as much as 35 percent of the total drag of the entire model. Obviously, the way to reduce this sort of parasite drag is to eliminate the launch lug by using a tower or piston launcher. But a suitable compromise is to locate the launch lug in the fin-body joint of one of the fins; this reduces the launch lug's parasite drag to only about 20 percent of the total drag of the model. The effort of relocating a launch lug to reduce drag by 15 percent is certainly worthwhile.

Induced drag is drag caused by lifting force, or drag *induced* by the lifting characteristics of a surface such as a model rocket's fins. A body tube and nose combination doesn't generate enough lift to really matter. In fact, Barrowman correctly ignores any nose–body lift in his CP calculation method because aerodynamic studies concerning the lift characteristics of bracing wires and struts on World War I biplanes indicated that such lift was negligible. The major lift-producing parts of a model rocket are its fins. They don't produce lift as you might normally consider it—lift *against* the force of gravity. They produce lift in the true and classic definition of the term: an aerodynamic force perpendicular to the surface. In model rocketry, this lift force is the aerodynamic stabilizing and restoring force we discussed in the last chapter.

Drag due to lift, or induced drag, is produced by any surface that generates lift. The reason for this is best understood by looking at Figure 9-4 where you can see that any surface generates drag as well as lift. But some surfaces can be designed to produce less induced drag because of two factors—their cross-sectional or airfoil shape, and the planform shape.

Without going into mathematical details, lift is generated by a surface at an angle of attack because there's high pressure on the surface most directly exposed to the oncoming airflow and low pressure on the other side. Even a flat plate generates lift when at an angle of attack. But specially designed shapes called airfoils generate more lift at lower angles of attack, are able to keep on generating lift at greater angles of attack, and possess a higher lift-to-drag ratio. (Model rocket fin airfoils are symmetrical airfoils and do not produce any lift at zero angle of attack.)

The most important source of induced drag is the airflow around the tip of the fin. If a fin is at an angle of attack and thereby generating lift to stabilize

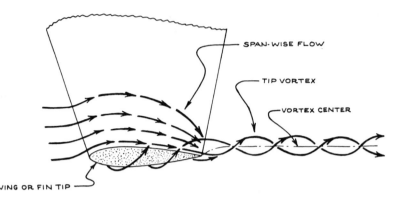

Figure 10-10 The creation of the tip vortex of a wing or fin generating lift. This is the source of induced drag.

the model and restore it to zero angle of attack, there is high pressure on one side of the fin and low pressure on the other. The high pressure tends to spill over the fin tip to relieve the low pressure on the other side. This creates a span-wise motion of air over the entire fin, more noticeably near the fin tip. Because the air is also flowing backward over the fin surface, the result is the creation of a corkscrew motion of the air, or vortex, that's shed from the tip of the fin. See Figure 10-10. It takes energy to create and maintain this vortex, and this energy loss shows up as induced drag. The smaller and weaker the tip vortex, the less the induced drag.

I hear a loud young voice say, "Well, just put another surface over the tip of the fin, making a tip plate to prevent the air from spilling over in the first place." This is certainly one solution to the problem, but one that requires very careful design trade-offs so that the reduction of induced drag isn't exceeded by the additional friction drag of the plate and the interference drag at the joint of the fin tip and the tip plate. It has paid off on some large aircraft where specially designed tip plates called "Whitcomb winglets" are used on high-performance wings.

But the simplest way to reduce the tip vortex and the induced drag is to shape the tips of the fins properly. Induced drag of several styles of fin tips is shown in Figure 10-11. Again, this data comes from wind tunnel tests made to determine the best shape for airplane wings. After all, a model rocket fin can also be considered as nothing more than a small wing. You might not suspect that an absolutely square fin tip would have the lowest induced drag of all the shapes tested, but it does. Several airplanes utilize this tip shape, most notably the very fast, highly streamlined, single-engined Mooney M-20 series. And a sharp tip also has low induced drag. This is also evident from looking carefully at many small, general aviation airplanes. Both tip

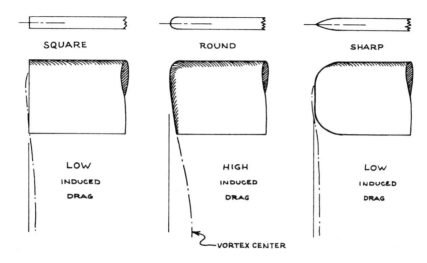

Figure 10-11 Induced drag of three simple fin tip shapes.

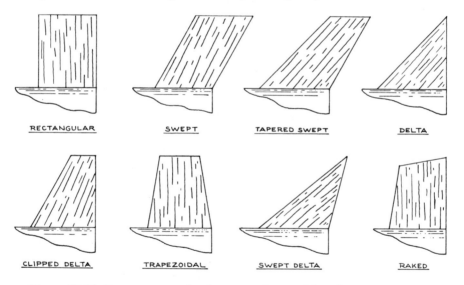

Figure 10-12 Some common fin shapes used in model rocketry.

shapes hinder the spillover of high pressure air to relieve the low pressure on the other side, thus reducing the size and strength of the tip vortex. It's also possible to install what is known as a conical tip, which is a tip that bends downward to create a semi-tip-plate; this works well on the asymmetrical airfoils of aircraft, but not on the symmetrical airfoils of model rockets.

The planform shape of a fin also has a great deal to do with its induced drag. Several common model rocket fin shapes or planforms are shown in Figure 10-12.

The swept-back fins that look so good on model rockets are actually the *worst* when it comes to induced drag. They can also exhibit what is known as "divergent flutter" which, in the case of large swept-back fins or the swept-back wings of high-speed jet aircraft, creates severe design problems if the wings and the airplane are to remain together.

The two best low-drag fin planforms for model rockets are the clipped delta and the elliptical. The clipped delta is shown in Figures 10-12 and 10-13. The elliptical planform looks something like the wing of the World War II British fighter plane, the Spitfire, which has perhaps the most aerodynamically perfect wing ever put on an airplane.

The difference in induced drag between the clipped delta and the elliptical planforms is about 1 percent in favor of the elliptical, but the clipped delta is easier to make. Therefore, I'm personally partial to the clipped delta and have designed many high-performance, record-setting model rockets using this planform.

The basic design parameters for the clipped delta planform are shown in Figure 10-13. It's based upon the diameter of the body tube. A clipped delta fin designed from these parameters tends to be slightly oversized for most model rockets, which gives you a margin of safety if you don't care to run the CP calculations. If you do and discover you can use a smaller fin, reduce the C_{root} and C_{tip} dimensions, not the S or span dimension. On the other hand, if you need more fin area, increase S, the span.

Now, how do you put all of this aerodynamic information together into a low-drag, high-performance model rocket design? A typical, idealized low-drag model rocket design is shown in Figure 10-14. It has an elliptical nose with a nose length-to-diameter ratio of 3. It has 3 fins with clipped delta planform. The launch lug is nestled in one of the well-filleted fin-body joints. The aft end of the body tube could not be boat tailed because the motor fits right into the body tube, so the aft end of the body is slightly rounded with fine sandpaper and a little bit of the motor casing extends from it. The extension of the motor casing beyond the aft end of the body tube has two functions; it makes it easy for you to grasp the motor casing to take it out after a flight, and the extension acts like an extension of the body tube to help reduce interference drag by smoothing the flow aft of the fins. The model ould have a shiny, mirrorlike surface with all balsa grain and tube joints filled. With this size model, it's extremely difficult to make it too heavy by filling and painting.

To prove that this idealized model is something that can be built, Figure 10-15 shows some of my competition and sporting designs.

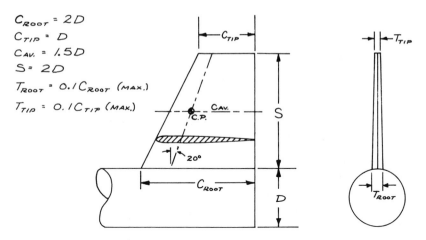

Figure 10-13 How to lay out a low-drag clipped-delta fin planform.

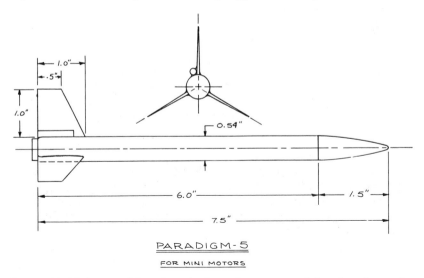

Figure 10-14 A typical low-drag, high-performance model rocket design for use with mini motors.

The drag coefficient (C_d) for an idealized model rocket such as Figure 10-14 has been calculated at 0.4 at a velocity of 200 feet per second; flight tests have indicated that the C_d can be as low as 0.25.

The drag coefficients (C_d) of the various parts of an average, well-designed model rocket have been calculated by Gregorek and Pantalos as follows:

Figure 10-15 Some of the author's low-drag, high-performance model rocket designs.

Nose and body pressure and friction drag = 0.2 (50 percent)
Base drag = 0.06 (15 percent)
Fin friction, induced, and pressure drag = 0.07 (17.5 percent)
Interference drag = 0.02 (5 percent)
Launch lug parasite drag = 0.05 (12,5 percent)

Again, these numbers have been generally confirmed by building models, flying them, and tracking them on six to ten flights. The actual altitude data is then compared against altitude data calculated with a computer program such as RASP-79E with various values of C_d. The closest comparison between actual altitude and computer altitude then provides you with a very good estimate of the C_d of the model.

The reason why there are so many different model rocket designs is that there are so many model rocket designers who handle the numerous design trade-offs differently. A model rocket designer must make technical compromises. A model rocket designer's handiwork is as distinctive as his signature because of the particular trade-offs and compromises he favors. You'll probably develop your own design signature, too.

There's still a lot left to be learned about the aerodynamics of model rockets. One of the hottest areas of controversy centers on aerodynamics, just as it does in model aircraft and even in full-size aircraft and space vehicles. All of the data hasn't come in yet, and a lot remains to be discovered. There's substantial guessing, many hypotheses unsupported by experimental data, and much data that hasn't been correlated into meaningful design rules. All of this indicates that model rocket aerodynamics is an area ripe for original research of the sort that really doesn't require an operating knowledge of advanced math, a supply of expensive equipment, precise measuring devices, or years of education and training. The aerodynamic area of model rocketry is full of fun and games. Just when you think you've got the whole answer, some other model rocketeer comes along with data that say otherwise—and the controversy is under way.

Don't think that controversy is bad for the hobby or for science and technology in general. At a 1957 space conference, the pioneer aerodynamicist and rocket expert, Dr. Theodore von Karman, was asked to sum up the meeting. He stood and told us, "A very fine meeting. Very well organized. Very well run. Excellent papers on very pertinent subjects. Good presentation. Solid data. *But no arguments!* Ladies and gentlemen, how can we possibly have progress without controversy?"

So, if you have the data to support your opinions, argue away. In the process, you'll have fun and learn a lot more. And someday, you may graduate to designing big things that fly.

11

Multistaged Model Rockets

Thus far, we've discussed only single-staged model rockets. There are two reasons for this. First, nearly everything that applies to single-staged model rockets also applies to multistaged model rockets—and sometimes even more so. Secondly, unless you understand the principles of single-staged model rockets that we've discussed already, you're going to have considerable difficulty with multistaged model rockets.

A multistaged model rocket permits you to make a significant improvement in the performance of a model rocket by increasing its total impulse and decreasing its final burnout weight by discarding as much unneeded weight as possible.

Technically, a multistaged model is one in which there are two or more motors that operate in flight with the expended motors and their airframes being discarded or jettisoned after they've reached burnout.

With a single-staged model, you can increase the burnout velocity by decreasing the weight of the model until the model is at the optimum weight—not so light that it starts behaving like a feather, and not so heavy that the motor thrust can't accelerate it enough. Or you can also accomplish this by super-careful streamlining and drag-reduction techniques. Or you can increase the total impulse of the motor.

However, there are limits to each of these methods beyond which you cannot go because you run up against technical, constructional, or operational limitations.

Multistaging offers a method of increasing burnout velocity without some of the disadvantages and limitations of single-staged models—but with some nasty little tricks of its own that you must watch out for!

The simplest form of staging is *series staging* shown in Figure 11-2. Essentially, the payload of the lower stage is itself a model rocket: the upper stage. A motor and its enclosing partial airframe—body tube, fins, and motor mount—make up the lower or first stage, often called a booster. The upper stage is a complete airframe, a single-staged model rocket. Ignition of the lower stage booster motor accelerates the entire multistaged model into the

168

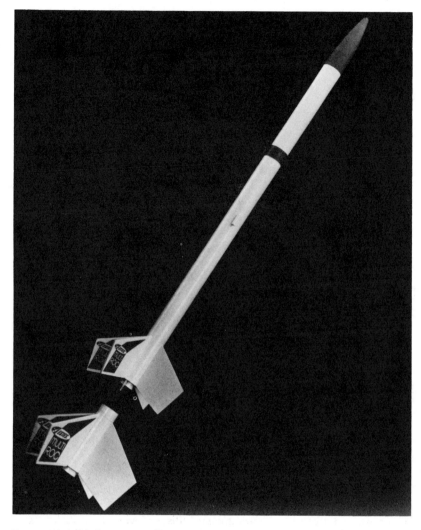

Figure 11-1 Multistaging offers a way to increase performance but adds complication to model rocket design and flying. (Estes Industries, Inc.)

air. At lower stage motor burnout, the booster assembly separates itself from the upper stage, and the motor of the upper stage is ignited. This adds the total impulse of the lower stage motor to that of the upper stage motor, which in turn adds the velocity imparted by the lower stage to the velocity that can be attained by the upper stage. Since we learned that the peak altitude is a function of the square of the burnout velocity, and since the burnout velocities

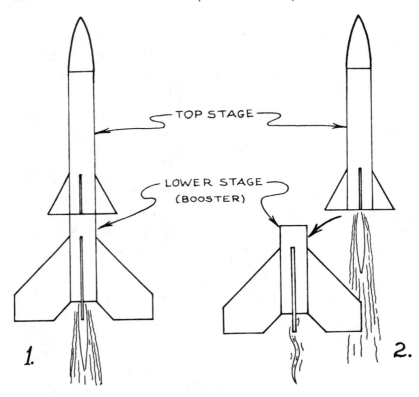

TOP STAGE

LOWER STAGE
(BOOSTER)

1.

2.

Figure 11-2 Series staging.

of the two stages are added together, you can quickly see that the peak altitude of a multistaged model can be very high indeed.

However, aerodynamic forces can also be very high.

In addition to increasing the total impulse of the model, series staging also decreases the final burnout weight of the model because the expended lower stage motor casing and airframe are jettisoned when their usefulness is at an end.

Separation of a series-staged model occurs with no apparent visible delay (actually, there is some, as we'll find out). Separation occurs when the lower stage has imparted its maximum velocity to the upper stage. Stage separation doesn't occur when the model reaches the maximum altitude to which the lower stage will carry it. Why is series staging done at maximum velocity rather than at maximum staging altitude?

An example will show you why. Let's use a hypothetical model rocket whose lower stage, without separation and ignition of the upper stage, can carry the entire model to a burnout velocity of 100 feet per second and a peak altitude of 250 feet as shown in Figure 11-3. Suppose that the upper stage all by itself can achieve a burnout velocity of 100 feet per second and a peak altitude of 250 feet.

If staging took place at the peak altitude of the lower stage, the peak altitudes would add together, giving a total peak altitude of 500 feet for the combination staged in this fashion.

However, if the upper stage is separated and ignited at the maximum velocity of the lower stage, the 100 feet per second velocity of the lower stage is added to the 100 feet per second of the upper stage, giving a final burnout velocity of 200 feet per second for the staged combination. Peak altitude is therefore four times what it would be at a burnout velocity of 100 feet per second, and the upper stage reaches a theoretical apogee of 1,000 feet.

In other words, a model rocket staged at maximum velocity of the booster will go roughly twice as high as one staged at the maximum altitude of the booster—neglecting aerodynamic drag, of course.

Another type of staging is called *parallel staging*, because the stages operate in parallel rather than in series. As shown in Figure 11-4, the parallel-staged model rocket leaves the launch pad with all motors of all stages operating. The staged motors/airframes are quite properly called boosters, and the upper stage is the core or the sustainer.

The NASA Space Shuttle and the old Air Force Atlas are parallel-staged rocket vehicles.

For parallel-staged model rockets, the booster motors are selected to have less duration than the sustainer or core motor. Perhaps the sustainer motor is a Type C6 while the booster motors are Type B8s. At booster motor burnout, the booster airframes are separated by air drag or other techniques, leaving the sustainer motor still thrusting and accelerating the model's core vehicle.

Parallel staging offers very high thrust and acceleration at lift-off. It also eliminates the problem of air-start of series-staged upper stage motors. However, ignition of multiple or clustered motors is also fraught with the possibilities of not igniting all motors on the pad. Modern low-current igniters have helped solve a lot of headaches in clustered ignition. For more details on clustered motor ignition, refer to the various technical reports published by the manufacturers that deal with cluster ignition of their particular motors.

For many years, it was believed that parallel staging wasn't practical for model rockets because of all the difficulties involved. Therefore, very little work was done on it. Pat Artis, of Ironton, Ohio, perfected parallel-staged model rockets and demonstrated them at the Seventh National Model Rocket Championships (NARAM-7) in 1965. By 1966, Artis had refined his designs to carry the booster pads up near the nose of the core model, thereby bringing the CG of the model well forward during the critical boost phase and eliminating the need for large fins to ensure stability during the boost phase. Artis dramatically showed that parallel-staged models had superior performance for

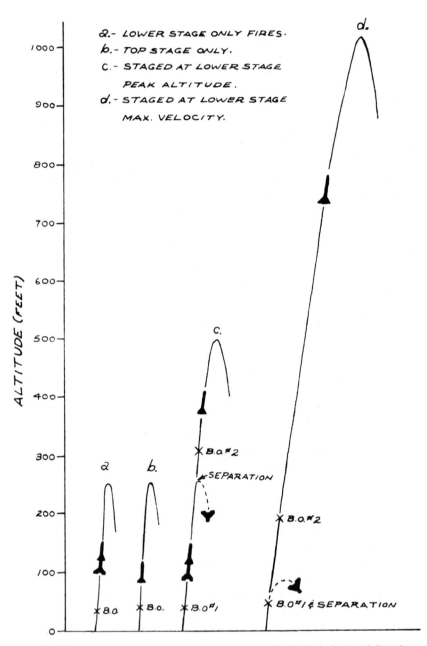

Figure 11-3 Series staging at various points in the flight of a model rocket.
(a) Lower stage only ignites. (b) Top stage only flies. (c) Model is staged at
peak altitude of lower stage. (d) Model is staged at lower stage burnout (B.O.)
point and therefore maximum stage velocity.

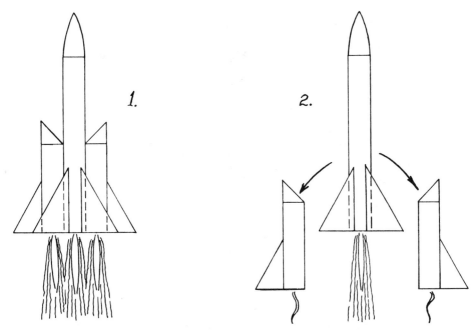

Figure 11-4 Parallel staging

Figure 11-5 Pat Artis (left) describes his parallel-staged model rocket to Cal Weiss of NASA. Note forward location of booster pods.

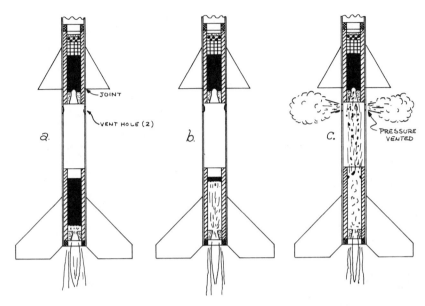

Figure 11-6 Cross section of a multistaged series-staged model rocket at various progressive points of lower stage operation and just before separation. Lower stage vent holes relieve internal pressure and prevent the lower stage from being separated before the upper stage motor ignites.

payload-carrying because of the high lift-off acceleration possible, thus enabling a heavy payload model to get airborne with an airspeed that would ensure stability.

However, since then, there hasn't been much activity in parallel staging. Perhaps this is because a lot of model rocketeers don't know about parallel staging.

Although there are some parallel-staged model rocket kits available, the series-staged model rocket is the most commonly flown type.

Success in flying a series-staged model rocket depends upon careful design, construction, and operation which, in turn, depends upon an understanding of what happens during a series-staged model's flight.

This is because air-start ignition of the upper stage motor can pose a problem if it isn't done carefully, especially in a model of your own design.

Ignition of the first stage motor of a series-staged model rocket is accomplished electrically in a manner identical to that of a single-staged model rocket, and also uses a launch pad. If the staged model is longer than 24 inches or weighs more than 6 ounces, a 3/16-inch diameter rod 5 to 6 feet long, a long tower, or a 5-foot C-rail should be used because lift-off acceleration will be low and the airspeed leaving the launcher may be low as well.

Figure 11-6 shows a cross-section of a typical two-staged series-staged model rocket at various times in its flight. Drawing A shows the model shortly after lift-off; the propellant in the lower stage motor is burning forward. Note that the lower stage airframe is nothing more than a hollow body tube with motor mounts installed and fins attached. The body tubes of the two stages are simply butted together; they're often kept in alignment either by the upper stage motor extending slightly into the lower stage booster as shown or, with models of diameters larger than the motors, by a paper collar extending from the booster stage up inside the upper stage to provide support and alignment. Thrust of the lower stage is transferred to the upper stage through the lower stage motor mount, lower stage body tube, and the butt joint between the two stages. The lower stage or booster motor should be firmly held in the booster body tube with a motor clip or with a paper ring *inside* the aft end of the booster body. (You have to insert the lower stage motor in from the front of the booster stage if you use the aft-ring motor retention method, so this is good only for short booster stages. If the model has a body diameter larger than that of the motors, an internal "stuffer tube" is used; this is essentially an extension of the lower stage motor mount tube forward to the upper stage motor. The stuffer tube prevents the expansion of the blow-through gases of the lower stage motor, as we'll see in a moment.

In Drawing B, the lower stage motor has almost reached burnout. There is only a thin disk of propellant left in the lower stage motor. Separation and staging are only a split second away.

In Drawing C the hot combustion gases under pressure have broken through the thin disk of propellant and are sending hot chunks of burning lower stage motor propellant forward. These pass up the nozzle of the upper stage motor, causing it to ignite. Once the upper stage has ignited, its jet pressurizes the lower stage body and blows it off the model. The motor clip or the aft-ring prevents the lower stage motor from being ejected from the booster airframe during this process. (If the lower stage motor casing is ejected and the booster airframe *isn't*, the jet from the upper stage undergoes an unusual phenomenon known an *over-expansion* or "Krushnic Effect" by blasting down through the lower stage body tube; this ruins the booster airframe and reduces the upper stage thrust to a couple of ounces, meaning that your model won't go very high but will probably deploy its recovery system before it lands.)

If staging is successful—and it can be 99.999999999 percent successful if everything's done right—the upper stage will then keep going up on its own, leaving the lower stage airframe to tumble back to the ground where it lands unharmed. Lower stages are deliberately designed to be unstable all by themselves so that they do tumble, and this takes some tricky figuring with the Barrowman Method.

The biggest problem with multistaging is ensuring the ignition and air start of the upper stage of a series-staged model. The problem manifests itself as follows: The model comes up on stage separation, and the blow-through feature separates the stages before the hot pieces of booster motor propellant have time to move forward and go up the nozzle of the upper stage motor to ignite it. When this happens, the upper stage doesn't ignite and therefore cannot eject the recovery device. Although the booster tumbles back safely, the upper stage performs what is known as a death dive.

It took a long time to find out exactly what goes on at stage separation and what to do to insure upper stage ignition.

First of all, various techniques were developed to delay the separation of the booster airframe until the upper stage had positive ignition and could blow the booster off.

Vernon D. Estes came up with the one that was most common for a long time. The model was designed so that the booster is short and the two motors butt together. The motor casings are secured together with a single wrap of cellophane tape. This keeps the motors together during staging until the upper stage ignition blows the motors apart at the taped joint. The drawback to this positive method of air-start ignition is the requirement that the motors butt together. This results in short, underdamped model designs that require very large booster and upper stage fin areas, often at the expense of drag, weight, and reduced dynamic stability.

The alternative was to tape the stage airframes together instead of the motors. This permits lengthening the booster stage airframe, resulting in smaller booster fins and better dynamic stability. However, sometimes the tape stays with the upper stage, creating excessive drag or nonsymmetrical drag.

A thorough investigation of what happens at staging was carried out by United States Navy Lieutenant Arthur H. "Trip" Barber III while he was still an undergraduate at MIT. He ran series-staged motors taped together in a static test stand. The data showed that the upper stage motor didn't ignite immediately after blow-through and burnout of the lower stage motor.

There was a delay of 0.03 second—thirty thousandths of a second or thirty milliseconds—between burnout/blow-through of the lower stage motor and the ignition of the upper stage motor. This was repeated, test after test. Thrust dropped to zero for thirty milliseconds at staging.

I came upon the solution to the series-staging air-start ignition problem as a result of a series of circumstances that are quite typical of how scientific and technical problems get solved. I'll go through it here not as an ego trip but to indicate to you how some of the everyday problems of science and technology are approached and whipped.

For almost a decade, I'd been thinking about this staging problem and making some flight tests and static tests, but I didn't have the high-resolution static test stands that were at Trip Barber's disposal at MIT. When I heard his data at a model rocket convention in San Jose, California, in 1977, I knew I was on the right track. Trip had provided the one missing piece of data I needed to proceed.

I'd been thinking conceptually about what really happens at staging. I'd hypothesized as follows: At the instant of breakthrough of the lower stage motor propellant disk, the lower stage body tube suddenly becomes full of two things: (1) hot combustion gas, and (2) burning pieces of fractured propellant from the lower stage motor. The hot combustion gas is less dense and has less total heat content than the burning pieces of propellant, and it moves forward through the booster body tube at the speed of sound, creating a rapid overpressure front known as a shock wave. The hot pieces of burning propellant are heavier; therefore, they move slower. The hot gas doesn't have enough heat evergy to ignite the upper stage motor; only the hot pieces of propellant can do that. But by the time the propellant pieces get to the upper end of the body tube, the shock wave of hot combustion gas probably has already blown the booster stage off the model.

Everybody who flew series-staged models knew that you had to keep the stages together until the ignition of the upper stage motor blew them apart. This was accomplished by taping motors and stage airframes together. But nobody knew why.

Trip Barber's discovery of the delay between booster motor burnout/ breakthrough and upper stage ignition meant that my hypothesis had some merit. It explained *why* motors or stages had to be taped together to delay their separation.

If the shock wave of hot combustion gas was blowing the stages apart prematurely, I reasoned that we had to get rid of the pressure of the hot combustion gases in the lower stage body tube.

So I vented the lower stage body tube by punching two 1/4-inch diameter holes on opposite sides of the tube just below where the nozzle of the upper stage motor would be. I used two vent holes because I didn't want the venting combustion gas to come out only one side of the model, thereby creating a side force like firing an attitude thruster; two vents on opposite sides of the body tube equalized the thrust forces from the venting gas.

The coupling between the two stages had to be tight enough to keep drag forces or any residual unvented gas pressure from blowing the booster stage off. If the coupling had been too tight, the jet exhaust from the upper stage motor would have vented through the holes until it literally burned the lower stage body tube away. So the coupling was made just tight enough that I could pick up the model by the aft end of the booster without having the stages separate.

I built a "standard" upper stage (several of them, JIC) and a series of boosters of varying lengths. Each of the initial test models was 0.976 inches in diameter. Later, I flew two-staged models with smaller and larger diameters to finally prove the method.

Over a period of several months, I flew over 100 two-staged flights with 100 percent success in upper stage ignition. Motors were separated by as much as 12 inches. At first, a lower stage booster airframe was good for only about six flights because the blast of the upper stage motor jet that blew the booster off the model also charred the inside of the booster body tube. So I cut a bunch of "liner papers" from ordinary 20-pound bond paper with dimensions such that I could roll one piece up and insert it into the booster body so that it provided an inner layer all the way around and extending from the booster motor mount right up to the lower edge of the vent holes. I used a new liner paper on each flight, and the booster airframes then lasted indefinitely.

Although this vented staging method appears to be similar to the former Centuri "pass-port" method, the vents on my models are open at all times during the flight. In the Centuri method, the booster body must move slightly to the rear to open the vents, which means that the booster has already started to separate from the upper stage before the combustion gas is vented, and there's a greater chance of the booster airframe being separated from the upper stage as a result.

Other model rocketeers have used the vent method, too, and it works for them as well. This means that the problem is basically solved. We can now achieve 100 percent air-start ignition of upper stage motors with boosters up to 12 inches long.

Since this is the first publication of the details of the vent method, you may not see it incorporated into kits or even other modeler's designs for some time. It takes time to incorporate changes into existing kits or to design, test, and market new kits.

When you're flying your first kit-built multistaged models, however, follow the instructions of the manufacturer relating to the staging of his kits.

You've probably noticed that the lower stage model rocket motors for both series- and parallel-staging are special. They have no time delay or ejection charges. Their NAR type numbers end in a dash-zero—B6-0, for example. This allows for staging at burnout of the lower stage motor.

Do not—I repeat, *do not*—use a standard motor with a time delay and ejection charge in a lower stage. The model will lift off normally, go through lower stage powered flight, then start to coast. Because a staged model is heavier than a single-staged model due to its additional stage airframe and motor, the model will coast upward and arc over. Because of the time delay, the lower stage motor will probably separate the stages with the model pointed *downward*. If the upper stage motor happens to ignite in spite of the head cap in the lower stage motor, the upper stage will be under thrust pointed *down* with the motor thrust assisted by gravity. It will come down *very* fast. There's no time for the recovery device to eject to slow the model, nor would it stay in one piece if it did deploy. The upper stage usually impacts under thrust. Very little usually remains of such "reentry models." And they are *exceedingly* dangerous.

SAFETY RULE: *Always* check carefully to make sure you've installed a dash-zero motor in all lower stages.

Although each lower stage motor must be a dash-zero type, the motor in the upper stage must have a time delay and ejection charge to deploy the recovery device. Because of the higher final burnout velocity of staged models, the upper stage has to coast for a longer period of time before all its momentum is converted to altitude. Therefore, upper stage motors must have a longer time delay to prevent deployment of the recovery device before peak altitude is reached. Long-delay motors are made and are specified for upper stages. When in doubt, however, use a shorter time delay on your first flights to prevent cliff-hangers where the model goes over peak altitude and deploys the recovery device on the way down.

As stated earlier, multistaged models weigh more at lift-off than single-staged models because of the additional weight of their lower stage airframes and motors. Therefore, with normal motors, multistaged models lift off slower and accelerate slower. Because of these lower launch velocities, good stability is important if a safe and predictable flight is to be achieved by a multistaged model. I've seen just about everything happen to a model rocket in flight, but the most frightening is a two-staged model that lifts off, becomes unstable, thrashes around in the air 25 feet above the launch pad, stages, and *then becomes stable*, usually pointed *down!* So check stability, *please!*

Multistaged models follow the same rules for stability as single-staged models. All of the stability requirements must be met. For a two-staged model, the CP and CG of the two stages together (in launch configuration with loaded motors in both stages) must be in a stable relationship. The CP and CG of the upper stage alone must be properly located. Thus, stability checks must be made for every flight combination of a multistaged model, i.e., in its various staged configurations and in its final top-stage configuration. For a three-staged model, three stability checks must be made—for all three stages together, for the top two stages together, and for the top stage all by itself. See Figure 11-7. Although you may occasionally get away with flying a new single-staged design without a stability check, you should *always* make stability checks for multistaged models. At the very least, subject the completed model to the simple swing test.

Because of their lower lift-off velocities and larger fin areas, multistaged models are *very* susceptible to weathercocking. Therefore, they should be launched in winds of less than 5 miles per hour. If in doubt about the wind, *don't launch* that multistager; wait until the wind dies down.

Lower stages are made with the same techniques as those used in building single-staged model rockets. In essence, the upper stages of a multistager become the nose of the lower stage. Recovery devices are not normally in-

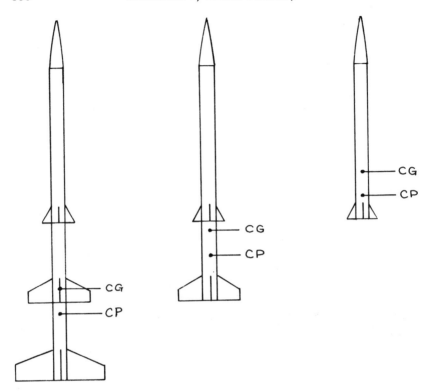

Figure 11-7 Stability checks on a three-staged model rocket. Each flight combination must be checked separately as shown.

stalled in lower stages (although I've done it) because of the need for the blow-through to ignite the next stage. A lower stage airframe, however, must be built in a more rugged manner than a simple single-stager. It will tumble during descent and land harder, which means that its fins—which are usually larger than a single-stage model's—have a greater chance to be broken off. Double-glue joints should be used throughout, and fin joints should be reinforced with tissue fillets.

Because lower stages don't deploy brightly colored recovery devices as single-stage or upper stage models do, they're more difficult to locate after they've landed, even though they probably won't go as far from the launcher. Therefore, lower stages should be painted a very bright fluorescent red or orange so they can be seen on the ground.

Couplings between stages should be designed and built so that the model will not bend or jackknife in the air at the location of the joint. Most kit models have this sort of solid joint, and there are numerous different kinds. Basically, the joint and coupling should permit lengthwise separation but prevent wobbling between the stages in the pitch–yaw axes. It's very frightening to have a multistaged model jackknife in midair and come apart because you're not quite sure when or if the upper stages are going to ignite and where they'll be pointed if they do.

Launch lugs may be placed anywhere on a multistaged model, but here's a tip: put the launch lug in a fin–body joint fillet only on an upper stage. This permits you to fly the top stage as a single-stager if you wish.

To achieve maximum possible stability during boost, don't align the fins of the various stages in a fore-and-aft direction. This puts the lower stage fins in the wake, downwash, and vortex pattern of the upper stage fins and greatly reduces their effectiveness—and you need all the effectiveness you can get from lower stage fins. Instead, *interdigitate* the fins. Put them out of line with one another as shown in Figure 11-8.

Upper stages can achieve some very high maximum velocities, sometimes well in excess of half the speed of sound (Mach 0.5). They should be well-built with exceptional care devoted to drag reduction if maximum performance is desired. They should also be built strongly because the drag force on an upper stage can exceed the weight of the model by a factor of ten or more. I've seen upper stages come completely apart after staging, leaving the sky full of fins and other parts. They have reached the legendary speed of balsa.

We've been speaking generally of two-staged models with a few references to three-stagers. However, the principles apply to three-stage models, too. However, don't try a four-staged or eight-staged model rocket. The United States Model Rocket Sporting Code limits a model rocket design to a maximum of three stages, although there is nothing in the NAR/HIA Safety Code or the Code for Unmanned Rockets NFPA 1122 about this. Why should you limit yourself to three stages?

Simply because the reliability of series-staged model rockets decreases according to the inverse of the square of the number of stages. Thus, a two-staged model is roughly one-fourth as reliable as a single-staged model, while a three-stage model is only one-ninth as reliable as a single-stager. This would make a four-stage model one-sixteenth as reliable.

But don't feel bad because a three-staged model may actually achieve a *lower* peak altitude than a two-staged model, even when the three-stager has more total impulse. This is because a three-stager is heavier at lift-off and because of air drag may actually have a final burnout velocity that's less than that of a two-stager. If a high burnout velocity is achieved, drag forces may rob the model of most of its altitude during the coasting phase.

Performance analyses conducted with a variation of the RASP-79A program have shown that a multistaged model rocket with more than three stages will not outperform a three-stager. Usually, it won't outperform a *single-stage* model rocket! You're wasting a lot of rocket power, which gets to be expensive.

Besides, a properly designed and successfully flown three-stage model

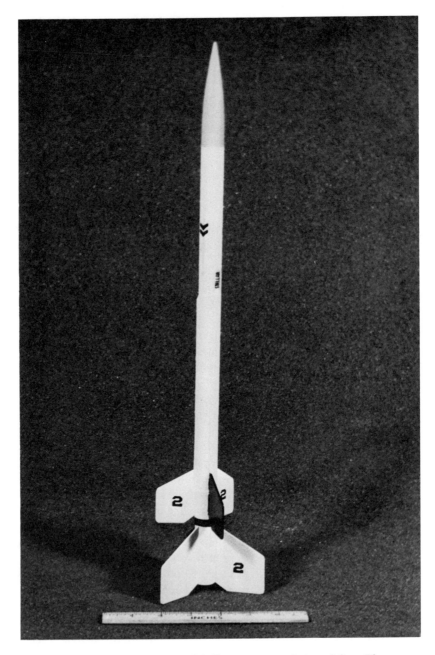

Figure 11-8 This two-staged model illustrates interdigitated fins. The upper stage fins are not aligned with the lower stage fins, thus making the lower stage fins more effective.

rocket will probably go completely out of sight. Then what are you going to do? You've probably lost it.

To boost a three-stager off the launch rod so it becomes airborne in a stable manner rather than lumbering and wobbling around the sky, use a high-thrust motor in the first stage. Such a motor is the Estes B8-0. It provides a high thrust that kicks the heavy model into the air with a higher launch velocity. Although the burning time is shorter than other motors, the B8-0 will probably have the model at a higher altitude at stage separation because it will produce a straight boost flight.

Multistaged model rockets are capable of exceptional altitude performance. But, because of the higher potential for something to go wrong, all safety precautions must be rigorously and very carefully followed. Tilt the launch rod a *few* degrees away from the vertical so that the model will land down range and away from people if it suffers a staging failure.

Because of the high performance and high-altitude capabilities of multistaged model rockets, you should always keep in mind the Federal Aviation Administration's regulations relating to model rockets. These are spelled out in detail in the Federal Air Regulations, Part 101, Subpart A, Part 101.1(a)(3)(ii), which is a federal law with teeth in it. Model rockets are exempt from FAA control if they weigh less than 16 ounces at lift off, contain less than 4 ounces of propellant, are made of frangible nonmetallic materials, and are flown in such a manner that they present no hazard to aircraft in flight. Model rockets exceeding these limitations are not considered to be model rockets by the NAR, HIA, NFPA, FAA, or any other organization; they're experimental or amateur rockets whose launch requires an FAA Air Traffic Control clearance and whose operation is against the law in the 49 states that have adopted or based their rocketry laws and regulations on the NFPA Code for Unmanned Rockets. It's easily possible to exceed the FAA and NFPA limits with multistaged models. So beware and be warned.

Multistaged models are fun. They're also difficult to build and fly. If anyone tells you that model rocketry is kid stuff and that you're playing with toys, he hasn't seen a multistaged model rocket climbing skyward at Mach 0.5.

12

Recovery Devices

Because recovery devices are required on all model rockets, the hobby is virtually free of hazards caused by freely falling objects that are large enough and heavy enough to do damage or cause injury when they return to the ground.

The rules of the NAR and the Federation Aeronautique Internationale (FAI) do not permit the jettisoning and free fall of any part of a model rocket, such as the motor casing, unless the falling part tumbles to slow its speed and uses a streamer or other readily visible surface that can be seen. The model rocket itself must be capable of more than a single flight and return to the ground so that no hazard is created. If you do everything right, follow all the rules, and read the instructions, you have a 99.999999999 percent chance of having your model rocket perform in this safe fashion. If something should go wrong, as it occasionally does, the design of the model and the materials used in its construction are intended to reduce the hazard to acceptable levels.

Although we've discussed recovery devices in general terms earlier, we'll now cover them in detail because there are all sorts of little hints, kinks, tips, and tricks about making and using them.

A number of highly successful and reliable recovery devices have been developed. The type of recovery device used in a given model rocket depends upon many factors—the gross weight of the model, the recovery weight of the model, the kind of payload, the size of the model, the anticipated weather conditions of the flight, etc.

All recovery devices are actuated or deployed by the ejection charge of the solid-propellant model rocket motor. This quick burst of gas from the ejection charge can be made to do a number of things to activate a recovery device.

NOSE-BLOW RECOVERY

I don't recall when the first nose-blow recovery system was used or who used it. It was definitely in use during the summer of 1958 during the flight testing of a large number of simple models near Denver, Colorado. It may have been first used by some model rocketeer who took the parachute out of his model to save time during flight preparations or to prevent the model from drifting away in a high wind.

Nose-blow is a very simple recovery method that derives from a similar technique used from 1946 onward at White Sands Proving Grounds, New Mixico, for aiding the recovery of German V-2 rockets. Normally, a 3-ton V-2 rocket falling from an altitude of 100 miles would dig a very large, deep hole in the desert, completely destroying all the scientific instruments aboard as well as the rocket itself. One day a V-2 happened to come apart on the descending leg of its trajectory and tumbled to the ground. The astounded rocket scientists and engineers recovered most of their equipment intact. Thereafter, the flight safety officer merely activated the ring of explosives around the base of the nose during the descending leg of the flight. The V-2's streamlining and stability were destroyed by this action, and the separated pieces fell at a much lower terminal velocity.

In model rocketry, the same principle lies behind nose-blow recovery, but we do it a bit differently. The nose is tied to the rest of the model by the shock cord. When the nose is separated from the model by the ejection charge, the aerodynamics and stability of the model are greatly altered. The CP–CG relationship is changed so that the model becomes unstable and flutters end over end. With the nose tied to the model by the shock cord, the two parts land together. This eliminates the need to search for both pieces separately.

A typical nose-blow model is shown in Figure 12-2.

Although a nose-blow model falls slowly enough to catch with your bare hands, it can still land hard enough to break fins if the landing site is an asphalt, concrete, or hardpan area. Therefore, most nose-blow models are constructed very strongly. A nose-blow recovery is used only for models weighing less than 2 ounces (60 grams). It is also used for high-altitude models when you don't want them to drift into the next state. However, one occasionally sees nose-blow used on large models with a high area-to-weight ratio so that their terminal speed is almost as low as a falling feather. Nose-blow is almost never used when flying a fragile payload.

STREAMER RECOVERY

When a model rocketeer wishes to slow his model a bit more than is possible with nose-blow recovery and also wishes to see it more clearly against the sky and on the ground after landing, he adds a streamer to the nose-blow recovery

Figure 12-1 Model rockets use recovery devices such as a parachute to lower them safely and gently back to the ground, provided they miss the rocket-eating trees!

model, creating a streamer recovery model. See Figure 12-3. Or he may build it from the start as a streamer recovery model.

A streamer is a brightly colored strip of plastic film or crepe paper attached to the shock cord or the nose base. The streamer is attached only at one end, although it's sometimes attached in its middle. Streamer dimensions vary between 1 inch and 4 inches in width and 12 to 48 inches in length. Many

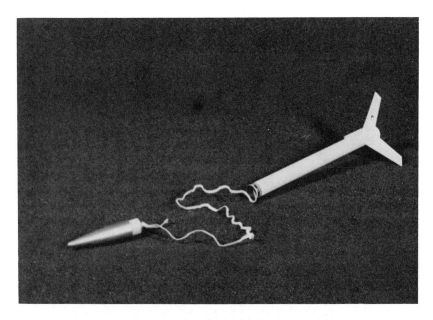

Figure 12-2 A nose-blow model rocket uses the ejection charge of the motor to separate the nose from the body, destroying the stability of the model. The nose is tied to the body by the shock cord.

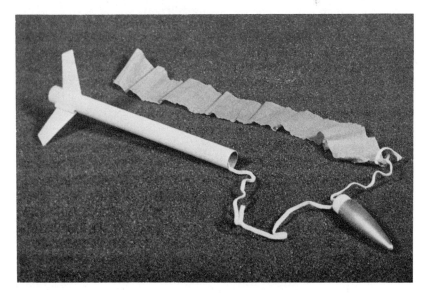

Figure 12-3 A streamer-recovery model rocket. The streamer may be attached by a separate line to the nose base or may be tied to the shock cord near the nose.

modelers, myself included, prefer to use orange crepe paper instead of plastic because crepe paper doesn't get stiff in cold weather, seems easier to fold and roll, does not burn or melt if some ejection charge gas seeps around the recovery wadding, and offers good drag characteristics.

Some work done by Trip Barber and others at the MIT Rocket Society indicates there's an optimum streamer size for every model rocket design. Generally, the optimum streamer size to obtain the slowest descent rate has a length-to-width ratio of 10. In other words, a streamer 1 inch wide should be at least 10 inches long, and a 2-inch-wide streamer should be at least 20 inches long. According to Barber's tests, one doesn't gain anything by going to a length-to-width ratio of more than 10 because a longer streamer actually begins to stream rather than flutter. To slow the model down, the streamer must flutter and flap back and forth.

Barber's work also indicates that it's possible to match the streamer dimensions to a given model rocket design to achieve the lowest possible descent rate and therefore the longest possible flight duration time. For this reason, the NAR has placed no limits on streamer size or material and has not attempted to adopt a "standard streamer" for its Streamer Duration Competition. This category of competition was originally intended as a simplified achieved-altitude contest for beginners and for "fun" meets. However, like other kinds of competition, it quickly evolved into a highly sophisticated contest category with its own strategy and tactics. In spite of this, however, Streamer Duration remains the simplest of all competition categories to fly.

A streamer must be protected from the hot gas of the model rocket motor's ejection charge. A plastic streamer can be melted and a crepe paper streamer can be burned by the ejection charge gas. Protection is simple and easy: Stuff a small wad of loosely packed flameproof tissue down into the body tube and put the rolled-up streamer in on top of it. The wadding blocks the ejection charge gas from the streamer and also acts like a piston to help eject the streamer. The wadding is ejected from the model after the streamer comes out. The wadding should *always* be flameproof material so that there's absolutely no chance it may be ignited by the ejection charge gas and fall smoldering to the ground where it could start a fire in dry grass.

A streamer is packed into a model by first rolling it tightly into a long cylinder. Often a single line of string or thread is used to attach the streamer to the base of the nose. This string can also be wrapped around the rolled streamer. The streamer should slide easily into the body tube so that it cannot be jammed when it's ejected later. It should be capable of being ejected easily by the ejection charge. If you wad it in, the ejection charge may not be able to move it. And don't push the streamer down into the front of the body tube more than about an inch, leaving just enough room to put the shock cord in atop it and place the nose on the tube; you want all the weight you can get up front to improve the CG–CP relationship.

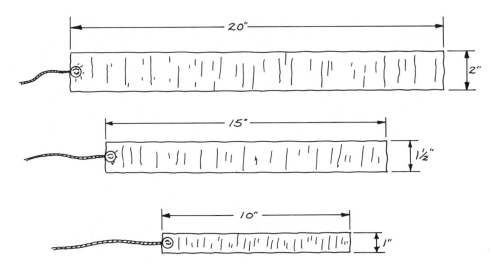

Figure 12-4 The best streamer configuration is a length-to-width ratio of 10 to 1 as shown.

A streamer recovery device will cause the model rocket to drift with the wind, and it'll go a great deal farther than a nose-blow recovery model. Often the model hangs up in a tree. In fact, it's a foregone conclusion among experienced model rocketeers that a model rocket is certain to land in the top branches of any tree within sight of the launch area at some time during the day's flying activities.

Most Streamer Duration competition models have the shock cord attached externally at the CG of the body/fin/motor casing, and the nose attached to the shock cord near the end where the streamer is attached. The shock cord lies along the external surface of the model until recovery deployment. This is to make the nose-less model descend in a horizontal position where it offers the greatest air drag, thereby decreasing the descent rate and increasing the flight duration time. In these competition models, the shock cord is an 18-inch to 24-inch length of nonelastic carpet thread, and there may be a notch in the nose shoulder to clear the shock cord as it emerges from the front of the body tube and lies back along the external surface of the body tube. The competition has gotten so sophisticated on the national and international level that many competition modelers use a wide variety of streamer materials such as tracing tissue, Mylar® plastic film, or bond paper; many competitors change the streamer for each flight since they've discovered that a "used" streamer may deploy with less drag than a fresh one. This only goes to prove that even a competition category that starts out as a simple beginner's altitude contest can evolve over a period of only a few years

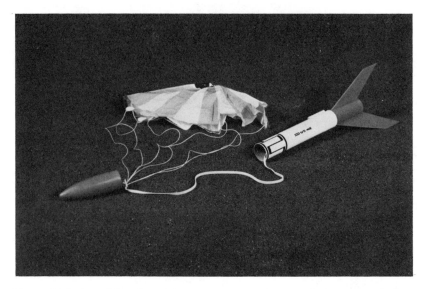

Figure 12-5 A model rocket with a recovery parachute made from polyethylene film and using twine or thread for shroud lines.

into a very complex event with highly specialized models and unique techniques and methods for flying it. But this doesn't keep Streamer Duration from being an excellent contest for beginners or for fun.

PARACHUTE RECOVERY

Parachute recovery of rockets is an old and established art. It may have been used by the Chinese, the inventors of the parachute itself, in their early fireworks rockets. But the first reported use in the literature is a description of a fireworks demonstration conducted by the Ruggieri brothers in Paris shortly after the French Revolution. (The Ruggieri company is *still* in the fireworks business and is one of the best in the world.) Dr. Robert H. Goddard successfully recovered many of his rockets with parachutes, and early experiments were made by the American Rocket Society (now the American Institute of Aeronautics and Astronautics) in the 1930s using parachute recovery. Some of these test rockets had replaceable motors like today's model rockets. Many modern research rockets regularly use parachutes for the recovery of the payload of even the entire rocket vehicle. Each of the two 82-ton expended Solid Rocket Boosters of the Space Shuttle is recovered by the use of a 54-foot diameter drogue parachute and three main parachutes, each 115 feet in diameter.

Recovery parachutes were used in the first model rockets made by Orville H. Carlisle in 1954. Since then, much has been learned about model rocket parachutes and many new techniques have been developed. A great deal of serious research work has been done concerning model rocket parachute recovery techniques.

A simple parachute may be added to a nose-blow model rocket or substituted for a streamer.

It's surprising that a workable model rocket parachute is so simple and so reliable. In fact, for many years, model rocket parachute recovery reliability was far greater than that for full-sized rocketsondes and research rockets.

Model rocket parachutes are commonly made from thin polyethylene plastic sheet or film ranging in thickness from 0.00025 inch to 0.001 inch. Model rocket manufacturers sell parachute kits that include brightly colored parachute canopies of various sizes and shapes. These colored canopies are easier to see against the sky than parachutes which are homemade from transparent dry cleaner's clothing bags.

Some high-performance competition parachutes are made from very thin and very strong plastic film called Mylar®. Such thin parachutes are also used when the storage space inside the model is small. Mylar® film is available in aluminized form and is superior for parachute canopies because it glints and reflects in the sunlight and can be seen for very long distances through heavy haze. Mylar® parachutes also have a "memory;" they "remember" the flat condition in which the plastic film was originally laid down in the factory, and they tend to open up and resume this original flat condition even at very low temperatures.

Ordinary polyethylene parachutes get very stiff at temperatures below 40°F. Under low-temperature conditions encountered in wintertime flying, these polyethylene parachutes often fail to open, resulting in a recovery device known as a "plastic wad" which doesn't have a very high drag coefficient.

Parachutes used in model rocketry are simple flat sheets of material. Therefore, they're technically not parachutes, but "parasheets." A true parachute isn't made from a flat piece of material and cannot be laid out flat on the ground; it's made from wedge-shaped gores of material that are sewn or fastened together to make a canopy that's hemispherical or semihemispherical in shape when inflated, as shown in Figure 12-7.

There's a parasheet that approaches a true hemispherical parachute in shape and drag characteristics. It's known as a "gathered parasheet." It's made as shown in Figure 12-8 by taking a flat piece of material, gathering the corners, and looping the shroud lines around each corner.

Although model rockets actually use parasheets, it's simpler to call them parachutes, and that's what we'll do from here on.

Shroud lines are normally made from cotton thread or carpet thread which is heavier. The shroud lines are usually attached to the skirt of the canopy by strips of tape or adhesive paper dots. Although most model rocket

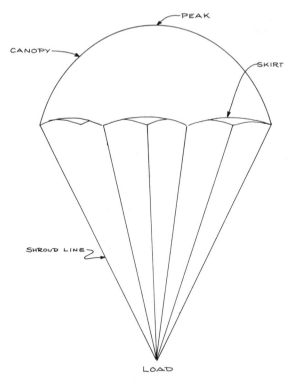

Figure 12-6 The parts of a parachute.

manufacturers supply tape dots or strips with their parachute kits for this purpose, sometimes better results can be obtained by using readily available tapes such as 3M's Scotch 810 Magic Transparent Tape® or, for cold-weather flying, one of the tapes designed for use in a food freezer. These tapes work well over a wide range of temperatures, don't pull off as easily as some tape dots often will, and don't lose their holding power with age.

To prevent the shroud lines from pulling away underneath the tape, loop them under the tape or knot them in their end, as shown in Figure 12-9.

A parachute made more than six months ago should be carefully checked and the shroud line tapes replaced with new ones if the tape fails in a hand-held "pop" opening test where you grab the load end of the shroud lines and snap the parachute open by hand. If shroud lines come loose during this test, more of them will come loose in flight, leaving you with a parachute that acts like a streamer.

You should try to make the shroud lines all the same length because

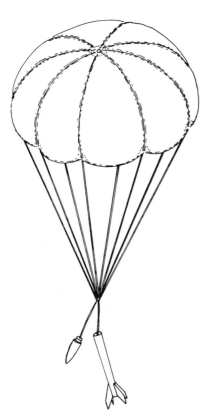

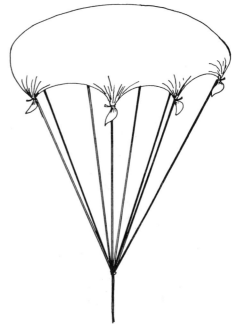

Figure 12-8 A gathered parasheet approximates the drag characteristics of a hemispherical parachute and is made from a flat piece of material gathered where the shroud lines are attached.

Figure 12-7 A true parachute is made with "gores" of material joined together to create a hemispherical shape when inflated.

this results in a symmetrical parachute that won't slip sideways in flight and that will have a lower descent rate.

There's little data available on the best shroud line length for a given parachute size. My experience indicates that the shroud lines should never be shorter than the major dimension of the parachute canopy. Thus, a 12-inch parachute should have shroud lines 12 linches long. Actually, I have been using shroud lines at least 1.5 times the major dimension of the parachute, and I've managed to win a few Parachute Duration contests and lose even more models to thermals or rising air currents where they've disappeared —*going up.*

A good research project would be: How is the descent or sink rate of a parachute affected by the length of the shroud lines? Make up several parachutes with different length shroud lines. Drop them from known heights

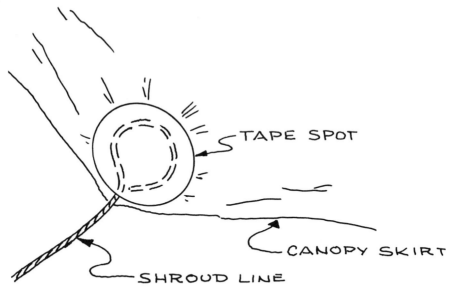

TAPE SPOT

CANOPY SKIRT

SHROUD LINE

Figure 12-9 Shroud lines are usually looped under the tape that attaches them to the parachute skirt. This prevents them from pulling out from under the tape when the parachute pops open.

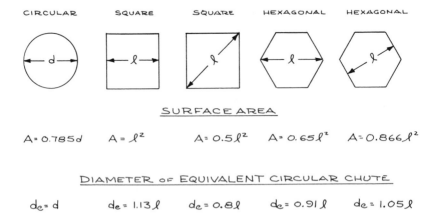

CIRCULAR	SQUARE	SQUARE	HEXAGONAL	HEXAGONAL

SURFACE AREA

$A = 0.785d$ $A = \ell^2$ $A = 0.5\ell^2$ $A = 0.65\ell^2$ $A = 0.866\ell^2$

DIAMETER OF EQUIVALENT CIRCULAR CHUTE

$d_e = d$ $d_e = 1.13\ell$ $d_e = 0.8\ell$ $d_e = 0.91\ell$ $d_e = 1.05\ell$

Figure 12-10 Some of the basic parachute shapes tested by Kratzer *et al* at the University of Maryland.

indoors in calm air. The farther you can drop them, the better the data. Time their descents to the floor. Do many, many drops—the more the better. It sounds like an easy project, and it is. But nobody's done it and *reported* it, and it needs to be done. Too many model rocketeers have become enamored by complex theoretical projects and have neglected research on the numerous everyday problems.

The desired number of shroud lines depends upon the size and shape of your parachute. There are only a few basic shapes, as shown in Figure 12-10. These are: round, square, hexagonal (six-sided), and octagonal (eight-sided). On circular 'chutes, six to eight shroud lines appears to be adequate. Most 'chutes have a shroud line at each corner—four for square 'chutes, six for hexagonal 'chutes, and eight for octagonal 'chutes.

Once the shroud lines have been attached to the canopy, bring the loose ends together in a knot and tie this to the base of the nose. Some modelers use a snap swivel of the sort used by fishermen; this allows them to attach and remove parachutes for different flight conditions.

To eliminate static electricity that may prevent a polyethylene 'chute from opening, and to "unstick" any shroud tape adhesive that may have gotten on the 'chute canopy, dust both sides of the 'chute with "parachute powder." The reason why the prep area smells so good at a Parachute Duration competition is that most model rocketeers use Johnson's Baby Powder® talcum for this purpose. This works fine, and its sweet smell tends to counteract the usual hydrogen–sulfide stink of old, used model rocket motors.

There's a wide variety of methods of folding and packing parachutes into model rockets. Some people merely stuff the 'chute in atop the wadding. Others develop weird methods of folding. I'm still using the same 'chute packing method taught to me by Orville H. Carlisle in 1957. It has worked perfectly for me over the years. This Carlisle Method is shown in Figure 12-11. Follow the steps shown, and you'll have a tightly rolled parachute cylinder that will easily slide down into the body tube and easily be ejected. When a 'chute packed by the Carlisle Method is ejected, it unrolls and deploys very quickly. I've heard 'chutes deploy and fill with air so quickly that they pop.

Although I always use flameproof wadding except where ejection baffles are built into the model (and these work very well, eliminating the need for recovery wadding), I take one additional step to be absolutely certain that the 'chute doesn't get scorched or spot-burned by any ejection charge gas that happens to leak past the wadding. It prevents the 'chute from getting ripped during ejection, too. I wrap the 'chute package in one square of flameproof wadding tissue. One layer is enough. It peels off quickly upon ejection from the model, and I've had it fail to do so only once or twice in several thousand flights.

How big a parachute is required? Well, how long do you want your model to stay aloft? Basically, the bigger the parachute, the longer the model

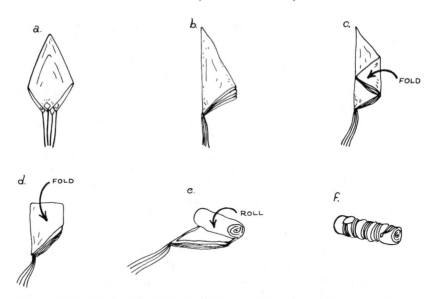

Figure 12-11 The Carlisle Method of folding, rolling, and packing a parachute.

will stay in the air. This general observation is based on the fact that a parachute produces a lot of drag. That's its function. Therefore, it must obey the drag equation:

$$D = 0.5pV^2C_dA$$

where A = the area of the parachute canopy.

Data from aeronautical engineering texts indicate that the C_d is about 1.5 while that of a parasheet is roughly 0.75.

Although model rocketeers knew about this equation and knew that it applied to parachutes as well, we really didn't pay too much attention to what it was telling us. We went by the rule of thumb that the bigger the 'chute, the longer the flight—and the farther you had to chase the model on a windy day.

Numbers were brought into the picture through a sophisticated series of parachute drop tests carried out in 1970 by Carl Kratzer, Bruce Blackistone, and Larry Lyons and reported by Doug Malewicki. They made a number of timed drops with various kinds of parachutes and standard loads from a 90-foot platform inside the Cole Fieldhouse of the University of Maryland. This provided them with a controlled environment of still air at reasonably constant temperature. The 'chutes were circular, square, hexagonal, and octagonal. The first test series were made of 0.00075-inch-thick polyethylene 'chutes from Estes Industries, Inc. Sizes were 8 inches, 12 inches, 18 inches, and 24 inches. The second series comprised parasheets made from half-mil (0.0005-

inch thick) transparent polyethylene from dry cleaner's clothing bags. This second series of half-mil 'chutes were made with diameters of 8 inches, 12 inches, 18 inches, 24 inches, 30 inches, 36 inches, 42 inches, and 48 inches.

A total of 240 drop tests were conducted by Kratzer with Greg Jones as timer and Bruce Blakistone as the loyal, hardworking recovery man. The 'chutes were hoisted back to the ceiling platform after each drop by a high-speed parachute crane consisting of a deep-sea fishing rod, line, and reel. Each of the 20 parachutes was tested with four different payload weights, giving a total of 80 different test combinations.

These tests confirmed experimentally what many model rocketeers already knew empirically from actual flying activities. However, some interesting new data was generated. For example:

1. The 8-inch 'chutes turned out to be "drogue 'chutes." That is, they would act to stabilize the falling payload, but didn't reduce its drop speed according to the drag equation. In some cases, these small chutes didn't open fully.
2. The performances of the hexagonal 'chutes and circular 'chutes were nearly identical.
3. Square parachutes drifted least.
4. The second series (cleaner bag 'chutes) opened easier and more completely, were approximate to the true hemispherical parachute shape, and drifted more.
5. The square, nondrifting 'chutes appeared to follow the drag equation very closely and had a calculated C_d of 1.0 in neat accordance with theory.
6. However, as any competition modeler will tell you, the big half-mil cleaner bag 'chutes performed best. Their calculated C_d was as high as 2.25. Since this is an "absurdly high value" according to Malewicki, he goes on to state, "It tells us that the 'chute is gliding and generating lift in addition to drag."

To see how a parachute can generate lift, look at Figure 12-12. Remember that the big half-mil cleaner bag 'chutes drifted the most in the drop tests. This sideways motion generated lift as shown in Figure 12-12.

This leads at once to some unanswered questions. Can a model rocket parachute be designed and built in the same manner as a Para-Commander or other full-scale skydiving sporting parachute? Can we build and use models of some of the sporting parachutes for model rocket recovery? Can the performance of a parachute be improved by deliberately making the shroud line lengths unequal to produce an eccentrically located load and therefore an induced angle of attack? Could the performance be improved by cutting a hole in one side of the canopy and asymmetrically venting the parachute? If so, how big a hole, where should it be located, and what shape should it have for best results? What is the actual effect of a vent cut out of the peak

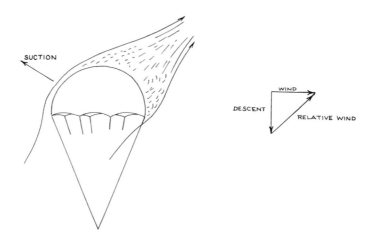

Figure 12-12 A parachute that moves sideways in the airstream will generate lift as shown.

of the parachute canopy? Does it really stabilize the parachute, and does it really improve the performance?

The answers to the above questions are identical: *We don't know yet because no model rocketeers have conducted and/or reported the results of experiments conducted under carefully controlled conditions such as Kratzer, Jones, and Blackistone did at the University of Maryland.*

It's very surprising that this sort of information isn't available in books and technical reports concerning full-sized parachutes. One would certainly think that this sort of research would have been done years ago. But it hasn't, and what little information exists on parachute performance usually isn't applicable to model rocketry for these reasons: (1) model rocket parachutes are very much smaller than ordinary parachutes and therefore are greatly influenced by scaling effects; and (2) model rocket parachutes are usually made from plastic film, which is nonporous, while all the available big parachute data relate to woven silk or nylon parachute canopies that let a small amount of air bleed through them.

So there's a lot of basic model rocketry parachute research left to be done. The experiments would be simple, inexpensive, and easily conducted. But if properly designed and carefully carried out they could give the answers to some important questions in model rocket parachute recovery technology.

The basic NAR rule that determines the parachute size that should be used for safe, gentle recovery is: 10 square centimeters of parachute area per gram of recovered weight. This works out to 44 square inches of area per ounce of weight in the English system. Therefore, a 12-inch parachute is suitable for models weighing up to 2 ounces, while a 24-inch parachute will handle models up to 8 ounces in weight.

Since no one has yet done any serious research and testing on the effects of the shape of the recovered model or payload dangling beneath the parachute and thereby affecting the airflow into the parachute canopy, we often run into some unusual situations. For example, I've had some models that had a faster sink rate with a 24-inch 'chute than with an 18-inch 'chute.

Remembering the basic drag equation, you can easily understand that the drag of a 'chute—and therefore the descent rate of any given model—increases directly as the canopy area increases (and as the square of the linear parachute dimensions). Descent speed is a direct function of the drag of a parachute. Doubling the linear dimensions of a parachute decreases the descent rate by one-fourth.

This is an important factor to keep in mind because there are times when you *don't* want a slow descent. The launch site may be surrounded by rocket-eating trees, or it may be small. Or the wind may be blowing strongly. A large parachute will cause a model to drift for a long distance. On the other hand, when a model is carrying a fragile or heavy payload, a slow descent may be mandatory to prevent damage to the payload upon landing. Therefore, the question of the best parachute size is subject to many compromises and trade-offs—there is no pat answer. This is why parachute duration competition is truly a sporting event that isn't simple and easy when conducted under either the NAR or the FAI rules.

OTHER RECOVERY DEVICES

In 1959, Vernon D. Estes invented and perfected a model rocket that would fly in one direction only: up. He utilized *tumble recovery* for the model. This recovery method makes use of the motor ejection charge to kick the motor rearward against a stop or otherwise redistributes the weight of the model. This moves the CP back behind the CG, and the model immediately becomes unstable. Thus, a tumble-recovery model rocket flies in a stable condition until the ejection charge changes the weight distribution which causes it to become unstable and tumble. Tumble recovery is limited to small, very lightweight model rockets. Estes Industries, Inc.'s first model rocket kit, the Astron Scout, was originally sold in 1961 and is still being sold at the time of this writing, making it the oldest model rocket kit in the world that is still commercially available. One of the first Astron Scout model rockets is on

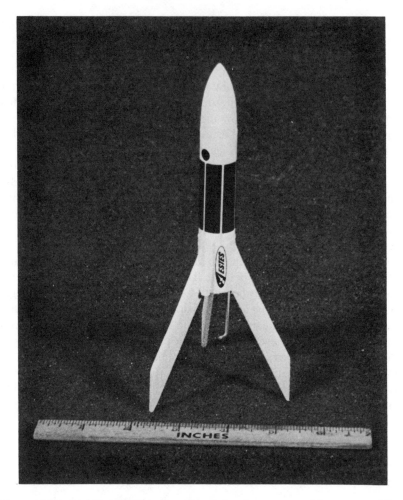

Figure 12-13 The Estes Astron Scout model rocket uses tumble recovery. The motor kicks back and brings the CG behind the CP. At this writing, this is the oldest continuously manufactured model rocket kit in the world, first put on the market in 1961. An original Astron Scout is on display at the National Air and Space Museum of the Smithsonian Institution.

permanent display at the National Air and Space Museum of the Smithsonian Institution in Washington, DC.

Featherweight recovery makes use of the familiar drag equation, too. Here, the motor casing of a very small, very lightweight model rocket is ejected from the model. The model then falls very slowly because of its high area-to-weight ratio. It's literally like a feather. This recovery method is also limited to very small and very lightweight model rockets. The NAR and the FAI don't permit freely falling motor casings in competition, so featherweight recovery isn't usually seen at contests. This method also suffers from the fact that once the motor ejects from the model very high in the sky, the tiny model becomes practically impossible to see. Most featherweight model rockets become one-flight models because they can't be found after they land.

Recovery by autorotation is commonly thought of, but rarely used in spite of the fact that there is an NAR competition category for "helicopter recovery duration." Some models eject or release helicopterlike rotor blades that spin and lower the model slowly to the ground. Other models allow their fins to cock to one side when the motor ejection charge goes off, thereby causing the model to spin, which slows the descent. These techniques are still in their infancy in model rocketry and are more difficult than the ubiquitous streamer and parachute recovery models.

And then there's *glide recovery* . . .

13

Glide Recovery

For decades, people have wanted to *fly* into space and *fly* back from space in winged, airplane-like vehicles. There were many different rocket airplanes; some of the first were designed and flown in Germany and the Soviet Union by people who later became the rocket engineers who built the first manned space craft. The Messerschmitt Me-163 "Komet" was a rocket-powered interceptor used by the Luftwaffe in World War II, but it was probably more dangerous to its pilots than to Allied bombers. The Soviets built, tested, and abandoned a rocket-propelled interceptor, the BI-2, during World War II. In the United States, high-speed aerodynamic research was conducted starting in the late 1940s with such rocket-powered manned aircraft as the Bell X-1, the Douglas D-558-2 "Skyrocket," and the North American X-15. Neil A. Armstrong, the first man to set foot on the moon, earned his astronaut's wings in the X-15 by flying it higher than 50 miles, and this was before he joined the Gemini and Apollo programs. Joe H. Engle, one of the pilots on the second Space Shuttle mission, also got his astronaut's wings in the X-15.

From the very beginning of model rocketry in 1957, we early enthusiasts believed it would be possible to make a model rocket that performs like today's Space Shuttle—vertical takeoff under rocket power and gliding return to a horizontal landing. But we didn't fully appreciate or understand the many problems involved in this unique marriage of model airplanes and model rockets.

There have been rocket-propelled model airplanes for a long time. The concept was first put forth by Werner von Siemens of Berlin, Germany, sometime between 1845 and 1855, but he didn't build and fly it. Probably the first were flown in Bucharest, Rumania, in 1902 by Dr. Henri Marie Coanda who, in 1962, watched me launch some modern glide-recovery model rockets. (History often has a strange thread running through it.) Ron Moulton, editor of the British publication *Aeromodeller* and a longtime supporter of "space modeling" in the Federation Aeronautique Internationale, attempted to build and fly glide-recovered aeromodels using skyrocket propulsion units in England in 1946; his efforts were frustrated by the lack of a small, reliable model rocket motor.

Rocket-propelled model airplanes became more common in about 1947 when the English Jetex rocket motors were introduced. However, Jetex motors produce very low thrust and very long durations; they are also heavier than model rocket motors with comparable total impulses. As a result, Jetex-powered model airplanes are just that—they fly under power in shallow, turning climbs with their wings always providing support against gravity. They're not normally capable of vertical takeoff.

Although model rocket motors of Type 1/2A through Type B don't have any more total impulse than the Jetex motors, they do have much greater thrust and lower weight. Early attempts to substitute Type A model rocket motors for Jetex motors in model airplanes resulted in some spectacular failures! The models designed for the gentle Jetex power couldn't withstand the high thrusts, high accelerations, and high airspeeds produced by model rocket motors. When the wings didn't peel off, the resulting violent and rapid loop produced a rather hard prang, to use the British terminology for a flight that terminates abruptly in a high closure rate between the aircraft and the ground, usually bending the aeroplane severely.

The first publicly demonstrated glide-recovery model rocket was demonstrated by Vernon D. Estes and John Schutz in 1961. It was flown at the Third National Model Rocket Championships (NARAM-3) in Denver, Colorado, in August of that year. Because the boosting portion of the model—the engine casing in this pioneer model—separated from the gliding portion, it was dubbed a "boost-glider," or simply a B/G for short. It had the normal three flight phases—vertical takeoff under rocket boost, vertical coasting flight to ejection charge activation, and recovery. But the recovery was as a glider.

The Estes–Schutz *original* boost-glider model—the very first one ever built—is now on permanent display in the National Air and Space Museum.

The Estes–Schutz B/G operated on a very simple principle: the motor ejection charge changed the aerodynamic configuration from that of a rocket to that of a glider. By today's standards, it wasn't a good model rocket, and it wasn't a good glider. But it worked. Basically, it wasn't a rocket-powered glider; it was a glide-recovered model rocket. The wings were large fins at the aft end of the model, as shown in Figure 13-3. When the motor ejection charge went off, the expended motor casing was ejected from the model. This not only reduced the weight of the model, but the action triggered a mechanical latch that released control surfaces on the trailing edges of the wings/fins, permitting little springs to force these "elevons" up. Technically, these elevons provided "negative camber" to convert the fins into wings. The model then entered a glide.

Once Estes and Schutz demonstrated a workable glide recovery model, the field of B/Gs literally exploded into creative development. Other model rocketeers, including many who couldn't get a B/G to work at all before, proceeded to get boost-glider models into the air and working. This is an interesting phenomenon that often occurs in technology. A thing may be

Figure 13-1 Glide recovery model rockets are among the most complex and advanced of all model rockets, combining the vertical takeoff of a model rocket with the slow gliding recovery of a winged shape. (Tony Medina)

considered impossible or extremely difficult until someone finally gets it to work, no matter how crude the workable solution may be. Once the break-through happens and people *know* it can be done because *somebody else did it*, developments follow very rapidly.

In the next five years, the skies were filled with wildly developed cut-and-try B/G configurations. Estes Industries, Inc. produced the Astron Space Plane B/G kit that worked in a similar fashion to the original Estes–Schutz B/G but which was carefully redesigned so that the average model rocketeer

Figure 13-2 The late Dr. Henri M. Coanda (in black hat) flew the first rocket-powered glider models in Bucharest, Rumania in 1902. He visited a model rocket launch in 1962 where the author (left) and A.W. Guill (right) explained modern glide-recovery techniques. (Lindsay Audin)

Figure 13-3 The first modern boost-glider designed and built by Vernon D. Estes and John Schutz in 1961. This is the original model now in the National Air and Space Museum of the Smithsonian Institution.

could build it and get it to work. The Astron Space Plane was a "flying wing" configuration that is rarely seen two decades later because better B/G designs have come along. Centuri Engineering Company soon followed with Lee Piester's Aero-Bat which is still an excellent B/G design, although the kit is no longer made. Paul Hans worked the bugs out of the canard configuration. Hunt Evans Johnsen of MIT built the first swing-wing variable-geometry B/G that looked more like the F-111 than the current Estes Scissors-wing Transport kit. Captain Bill Barnitz of the United States Air Force, an F-4 pilot, tackled the flex-wing configuration. All of us were reinventing the wheel, however, because we really had very little knowledge about why or how. We were engaged in empirical experimentation with little background in the theory of lifting surfaces and aircraft aerodynamics.

This was due to another common phenomenon in technology which can and should be avoided.

In spite of the fact that model rocketry owes a great deal to model aviation, people don't come into the hobby of model rocketry by way of model aviation; they usually come right into model rocketry without having previously built a flying model of any type, not even a simple balsa chuck glider. For many years, model rocketeers and model aviators didn't—and to some extent still don't—talk to one another. The model rocketeers thought the model aviators were hopelessly old-fashioned with their propeller-driven flying machines. On the other hand, model aviators considered model rocketeers to be hopeless pyromaniacs who indulged a fascination with fire and smoke while cluttering up the air with ugly little plastic parachutes. This lack of communication forced model rocketeers to go through a long period of trial-and-error experimentation with B/G models, making the same mistakes all over again that the model aviators had experienced years before. Model aviators could have explained a great deal about hand-launched and catapult-launched gliders to model rocketeers because a B/G turns out to be a rocket-launched version of a hand-launched glider in most cases.

Whenever communications between model rocketeers and model aviators were established, even briefly and almost accidentally, the results were spectacular. This should serve as a lesson to us all. A lack of communication, an inability or unwillingness to listen, and a failure to seek out all pertinent information can lead to a lack of progress in any field of human endeavor, not just model rocketry.

The second big breakthrough in B/G technology came about as a result of communication between model rocketry and model aviation. Larry Renger, a model aviator and then a senior at MIT in aeronautical engineering, took a basic hand-launched glider and put a model rocket motor up front, mounted on a pylon. It was a conventional wings-and-tail airplane configuration, something none of us had bothered to try because our early experiments had failed for many reasons. With the model rocket motor up front, the model was very

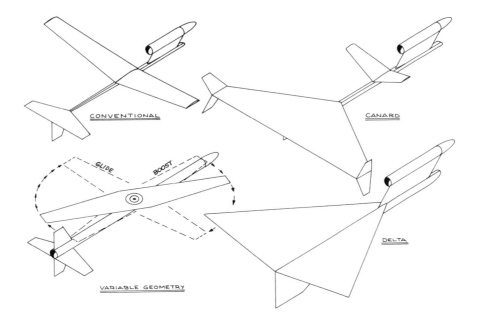

Figure 13-4 Basic glide-recovery model rocket types.

nose-heavy and became essentially a ballistic vehicle like a model rocket. When the ejection charge popped the motor casing out of the model, it removed weight from the front and shifted the CG rearward, reducing the gliding weight of the model at the same time. The glider then settled into a nice, low sink rate glide.

This was the famous Renger Sky Slash design—a rocket-powered glider, not a gliding model rocket. The emphasis was on boost-*glide* rather than the other way around, *boost*-glide, as in previous B/G designs.

Properly trimmed, a Renger Sky Slash would outglide almost any other type of B/G then flying, even though the original Renger design had been simplified and changed in the plans that Estes Industries published. The Sky Slash design dominated boost-glider duration competition here and abroad for over a decade. Only recently has it been challenged by the developments in flex-wing technology revealed by the Bulgarians at the 1980 World Space Modeling Championships held at Lakehurst, New Jersey, where the Bulgarian B/G team walked away with the gold team medal plus gold and bronze medals for individual first and third places.

The ultimate refinement of the front-motored conventional configuration B/G came with the development of the pop pod to eliminate the potential

Figure 13-5 The original front-motored rocket-boosted glider design, the Renger Sky Slash, as built from the Estes plans.

safety problems associated with ejecting an empty motor casing. There is some question about who actually invented the pop pod. I was flying B/G models with pop pods in April 1965, and I subsequently learned that Larry Renger had designed the FlaminGo with a pop pod at about the same time. Here is an example of parallel technological development because it was obvious that something like a pop pod would be required in order to conform to the competition rules. Besides, it would greatly improve the glide performance of a B/G if you could jettison *everything* that contributed weight or drag to the glider portion of the model.

The boost-glider concept permits the model to drop its propulsion portion just as the Space Shuttle drops its SRBs and External Tank. But even today, professional rocket engineers are at work on future shuttle concepts that don't drop *anything* and use up only rocket propellants during the mission or flight.

About 1971, model rocketeers also began to think about the challenge of glide-recovery models that didn't drop *anything* in flight.

The result was the *rocket glider*, the RG, which differs from the B/G in that the B/G is permitted to jettison its propulsion portion while an RG must go up like a rocket and glide back without separating *anything* in flight.

A rocket glider is truly a glide-recovered model rocket. It follows the basic single-staged model rocket concept: what goes up together must come down together in a safe manner.

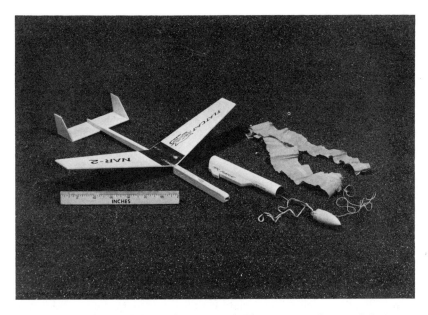

Figure 13-6 A boost-glider with a pop pod. The ejection charge of the motor kicks off the nose and ejects the pod streamer. Reaction kicks the pod backward, disengaging a hook on the pod pylon. This is the author's Flat Cat beginner's B/G model design.

The 1970s saw the long, painful development of RG technology just as the 1960s saw that of the B/G.

An RG model rocket is exceedingly difficult to build and fly, making it probably the most advanced area of model rocketry at the time of this writing.

Probably the most concentrated and thorough research and development in the RG field has been done by Howard R. Kuhn of Alexandria, Virginia.

There are three ways to make an RG work:

1. Use the ejection charge to cause a change of the CG of the model. This could be accomplished by shifting the position of the motor casing at the time of ejection charge activation, moving the motor rearward, catching it, and holding it there.
2. Use the ejection charge to change the aerodynamic configuration of the model. The ejection charge could be used to activate elevons or other surfaces, as in the early B/G models, or to actually change the relative positions of the aerodynamic surfaces of the model.
3. Use the ejection charge to cause a change that combines both of the above.

Kuhn methodically considered, designed, built, and tested dozens of RG designs to evaluate all three of the above solutions. He came to the conclusion that Solution Number 2 was the best. His design, the Buzzard, features a *movable wing* that slides back and forth on the fuselage. The Buzzard looks like a standard conventional configuration hand-launched glider or B/ G. During boost, the wing is held at the rear end of the fuselage next to the elevators and rudder by a thread that goes forward and passes through the forward power pod body in front of the motor. The Buzzard is launched like a model rocket and, with the wing all the way back at the end of the fuselage, it's like an ordinary model rocket with a pair of big fins on it. When the ejection charge goes off, the thread holding the wings is burned through. A rubber band pulls the wings forward on the fuselage up against a forward stop, and the Buzzard converts into a glider. The Buzzard can be built and flown by modelers with average experience, and it's available in kit form from Competition Model Rockets as the only sliding-wing rocket glider kit available at this time.

Careful design work by experienced model rocketeers such as Chris Flanagan and Jim Pommert has perfected RG designs based on Solution Number 1 with RG designs of conventional configuration having moving pods or moving motors to shift the center of gravity (CG) location or, much more difficult, models whose weight is shifted merely by the expenditure of propellant in the model rocket motor. This last design approach seems to result in models that start to level off and glide at or shortly after burnout, as you might expect.

RG is still a highly experimental area of model rocketry. In spite of Kuhn's outstanding empirical work in RG that is similar to Vernon Estes' development efforts on the Astron Space Plane BG kit in 1961, RG has not yet received the relentless scientific scrutiny given to the B/G field.

In fact, the whole field of glide recovery is one of the hottest in model rocketry. Rapid progress is being made in all areas. Nothing stands still. An entire book could now be devoted to glide recovery alone. Since we can't possibly cover all the details of this complex field, this chapter will have to serve primarily as an introduction to one of the most fascinating areas of model rocketry.

Much of the following information is applicable to both boost-gliders and rocket gliders. Some is pertinent only to boost-gliders because they've been around longer and we understand them better.

FLIGHT OF GLIDE-RECOVERY MODEL ROCKETS

Boost-gliders and rocket gliders may be considered either as model rockets with gliding recovery or as gliders with rocket boost. Depending upon where

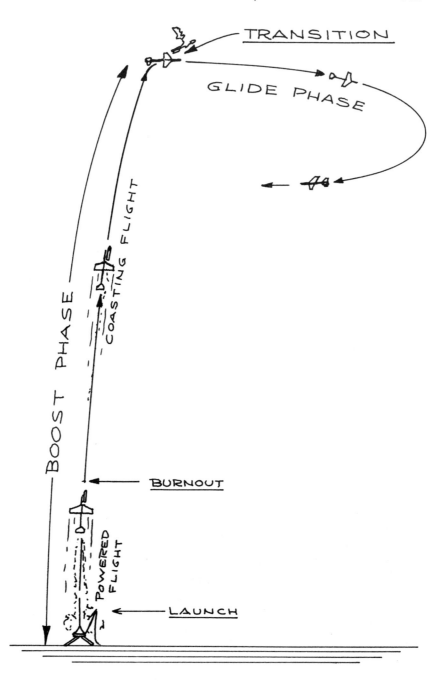

Figure 13-7 The basic flight phases of a glide-recovery model rocket.

you place the emphasis, they are *boost*-gliders or boost-*gliders*. The same holds true for rocket gliders.

The flight of a glide-recovery model rocket can be generally divided into two separate and distinct phases—the boost phase and the glide phase. The requirements for stability, structural integrity, and aerodynamics differ markedly from the boost phase to the glide phase. This simple fact means that a glide-recovery model rocket must always be a compromise between a model rocket and a glider. The sort of technical trade-offs made by the designer dictates which kind of glide-recovery model rocket the result will be.

The same simple fact also dictates that a change must take place for the model to make a successful transition between boost and glide phases. In the boost-glider, the booster airframe and motor casing are jettisoned, causing a change in CG location, gliding weight, and other factors. In a rocket glider, the ejection charge of the motor must shift the CG location or change the aerodynamic configuration, but there is no change in weight.

Boost phase is that portion of the flight during which the model performs like a ballistic vehicle—a model rocket—launched from a standard launch pad with electric ignition. It's propelled aloft in a near-vertical trajectory by the high thrust of the model rocket motor. It trades vertical velocity and momentum for altitude in the time-honored fashion of a model rocket during the operation of the motor's time delay. During boost phase, a glide-recovery model is not supported by lifting surfaces operating against gravity; whatever aerodynamic lifting surfaces are exposed to the airstream during boost must either have no effects upon the ballistic flight path of the model, or they must stabilize the model in the same manner as fins stabilize an ordinary model rocket.

The characteristics of boost phase include both high acceleration and high airspeeds. These can combine to produce a situation aptly described by the model rocket term introduced by the Czechoslovaks: striptease. This is an understandable word in many languages! Large wings rip off, and poorly made glue joints may turn loose. Therefore, glide-recovery models must be strong, and very good construction techniques must be used. The models usually have short, stubby wings in comparison to nonboosted high-performance model or full-sized sailplanes.

The glide phase has entirely different characteristics. The model in a glide is supported by the lifting force of its wings, which sustain it against gravity by the forward airspeed creating airflow over the wings. This airspeed is as low as 10 feet per second or less. The sink rate, or vertical descent rate, may be as low as 1 foot per second. Since the sink rate of a glider is directly related to the ratio between the lift and drag of the glider, very high lift-drag ratios on the order of ten or more are desirable, if you can obtain them.

As during boost phase, the gliding model must be stable in pitch, yaw, and roll. But the pitch and yaw axes can no longer be considered identical because most glide-recovery models are not symmetrical in the pitch and yaw axes. The glide-recovery model must have very high inherent stability because winds, gusts, and thermals (rising air columns and bubbles) will continually disturb its equilibrium during glide. If it pitches nose-up, the forces generated by its lifting/stabilizing surfaces must bring the nose down into an equilibrium or balanced gliding condition again. The same holds true for nose-down pitching. If the model rolls, it must right itself. If it yaws right or left, it must correct itself through proper design. Unlike an ordinary model rocket, a glide-recovery model during glide must always "know" which way is up, where the gravity vector force is.

A fin-stabilized model rocket is not a good glider, and a glider is not a good model rocket. Since the two flight phases are very different in their requirements for design and construction, compromises must be made. If maximum flight duration is the objective, it can be obtained in either of two ways. A modeler can strive for maximum altitude during boost, compromising good glide characteristics and obtaining long flight duration by simply having the model take a long time to descend from a very high altitude in spite of the fact that it may have a high sink rate. Or the glide-recovery model designer can work towards achieving maximum performance during the glide phase, cutting back on the altitude gained during boost in favor of low sink rate and other more favorable glide characteristics.

A great deal depends upon the weather at the flying site when you get ready to launch. Many modelers take two different glide-recovery models to a contest—a foul-weather model and a fair-weather model. A foul-weather model usually has very good boost characteristics and a very stable glide, while a fair-weather model is usually designed for absolute maximum glide performance in calm or nearly calm air.

Stability and Control of Glide-Recovery Models

Except for the radio-controlled glide-recovery model rockets being built and flown by advanced model rocketeers, small glide-recovery model rockets of the sort built by most modelers can't be controlled during their flight because they're too small to carry a pilot or even modern solid-stage radio control receivers, batteries, and servos. Therefore, a glide recovery model rocket must have inherently good stability and be able to maintain its equilibrium during flight because of this good stability.

Roll stability during boost may not be required. In fact, some models are deliberately rolled slowly during boost by means of offset pods or motor thrust lines. This causes the model to "screw" its way up into the sky. Basically, this averages out any small pitch or yaw motions that occur during boost phase and is considered a "quick and dirty fix" for any model rocket, glide-recovery or not, that may be *slightly* unstable or dynamically undamped. Too much roll during boost phase can be self-defeating, however, because it can greatly reduce the peak altitude attained.

During glide phase, roll stability can be obtained by using a dihedral angle on the wings as shown in Figure 13-8. This dihedral angle should be between 10 and 20 degrees per wing. It can be made up of compound dihedral angles—i.e., a straight center section with the tips turned up to the same location they'd have if the wing had straight dihedral. (This was the quick-fix used by aeronautical engineers on the F-4 "Phantom II" jet fighter plane; it needed more dihedral for roll stability, but adding straight dihedral to its wings would have required extensive redesign of the wing-fuselage structure.) The optimum dihedral for a wing is actually elliptical as shown in Figure 13-9.

Stability in the yaw axis is obtained by a combination of vertical stabilizers (which may have movable portions called "rudders"), wing sweep-back, and/or wing dihedral. Normally, adequate yaw stability can be obtained with a vertical stab area equal to 5 percent to 10 percent of the wing area, as we'll see later.

The biggest control and stability problem of all glide-recovery models in both boost and glide phases is in the pitch axis. See Figure 13-10.

During boost phase, an improperly designed, constructed, or balanced glide-recovery model—except for the flex-wing types that store the gliding wings folded inside a body tube or "parasite" types that are very small in comparison to the boost pod or vehicle—will pitch up into a loop or pitch down into an outside loop. Although these aerobatics are often spectacular, the model usually prangs. Or it loops so tightly that it doesn't gain sufficient altitude for a long glide. Or the Flight Safety Officer or Judge yells, "DQ!" That means the flight's been disqualified as unstable and/or unsafe.

The radius of the loop caused by pitching depends upon (1) the location of the CG, and (2) the wing-loading of the model (in ounces per square inch of wing area, for example). The farther aft the CG or the lower the wing loading, the smaller the loop radius. Therefore, the most desirable model characteristics during boost phase are *forward location of the CG* and *high wing loading.*

The CG characteristic is why a front-motored glide-recovery model performs better than a rear-motored one. The wing loading factor is the reason why a parasite glider—a gliding portion of a B/G that's hung on the side of a much larger booster vehicle such as the Estes Orbital Transport and therefore has a very high wing loading when attached to this booster—works so well during boost.

Improper balance or trim (location of the CG) during glide phase will cause the glide-recovery model to pitch up into a stall or dive into the ground. A badly trimmed glider may also get into increasingly steeper and more violent

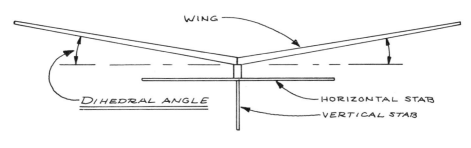

FRONT VIEW

Figure 13-8 Dihedral angle of the wings of a B/G as seen from the front.

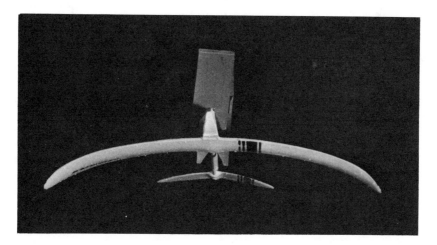

Figure 13-9 Elliptical dihedral of a wing.

stalls, ending up in a spiral dive to the ground.

Stalls are caused by the CG being too far aft—a tail-heavy model, in other words. At its worst, this aft CG condition leads to sharper and deeper stalls until the model goes into a spiral dive from which it never recovers.

With a nose-heavy model where the CG is too far forward, there's very little gliding flight at all. The model simply goes into a dive and prangs.

Control and stability of a glide-recovery model during the boost and glide phases are a matter of *balancing* the forces generated by the various surfaces lying in the pitch plane—the wings and the horizontal stabilizer (if any).

As shown in Figure 13-11, the CP of a flat plate or symmetrical airfoil, including symmetrical model rocket fins, is always located at a distance of

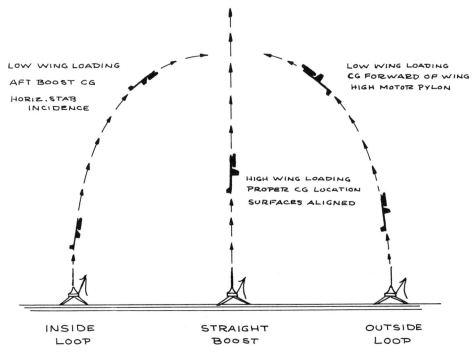

LOW WING LOADING

AFT BOOST CG

HORIZ.STAB
INCIDENCE

LOW WING LOADING
CG FORWARD OF WING
HIGH MOTOR PYLON

HIGH WING LOADING
PROPER CG LOCATION
SURFACES ALIGNED

INSIDE
LOOP

STRAIGHT
BOOST

OUTSIDE
LOOP

Figure 13-10 Pitch axis stability problems of glide-recovery model rockets in boost phase.

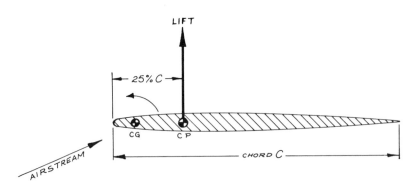

LIFT

25% C

CG   CP

AIRSTREAM

CHORD C

Figure 13-11 Cross section of a symmetrical airfoil with CP located at 25% of the chord and the CG located ahead of the CP. Pitching moment will always be in the direction of the uncoming airstream, making the airfoil always point into the airstream. This is a stable condition for boost phase but will not produce a glide.

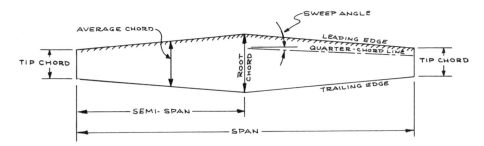

Figure 13-12 Definitions of the dimensions of a wing.

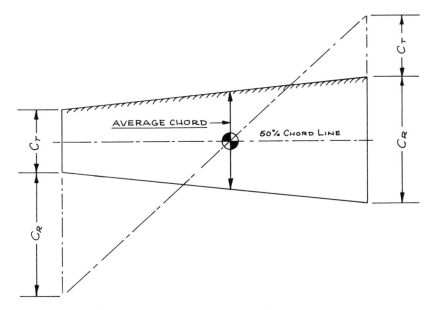

Figure 13-13 How to locate the average chord by geometrical means.

25 percent aft of the leading edge of the average chord of the surface.

Now, what do we mean by those words?

Figure 13-12 tells you. The chord of a wing, stab, or fin is its fore-and-aft dimension parallel to the body tube or fuselage. Unless a wing has a rectangular planform (a "Hershey-bar wing") with the same chord from root to tip, the average chord is the sum of the root chord plus the tip chord, the sum being divided by two. You can precisely locate the average chord by geometrical construction as shown in Figure 13-13.

If we place the CG ahead of the center of pressure (CP) and put the airfoil in a glide condition at an angle of attack as shown in Figure 13-11, the lift force concentrated at the CP will tend to make the wing rotate and pitch down or into the airstream, reducing the angle of attack to zero. With the CG ahead of the CP, the surface always seeks zero angle of attack. This is precisely the situation we want for the fins of a good fin-stabilized model rocket, but not for a glide-recovery model rocket during glide phase. This condition will simply make it dive right into the ground.

Therefore, to make a model glide with a CG forward of the wing's CP, we must have some mechanism that will bring the nose up when the model starts to dive and, conversely, bring the nose down when the model starts to climb and approach a stall. We must somehow obtain a *balance of forces*. Our glide-recovery model is just like a teeter-totter with its balance point being the CG. Anything that makes it want to dive should immediately produce a force to make it climb and bring it back to level flight again. And, conversely, anything that makes it want to climb should produce a pitch-down force that will bring the nose back to level flight.

In other words, the model must be stable in the pitch axis just as it's stable in the roll axis, thanks to wing dihedral, and in the yaw axis, thanks to the vertical stab, wing sweep, and wing dihedral. Any tendency of the model to wander away from its balanced, stable, level gliding flight condition must be counteracted by forces that will return it to this balanced gliding condition.

As you can see, this stable glide condition is radically different from the stability condition required during boost phase. Therefore, something must happen to the model to convert it from a ballistic, nose-heavy, fin-stabilized model into a gliding model. And this should logically occur at or near peak altitude achieved during the boost phase.

A glide-recovery model is designed and built so that the motor ejection charge action changes the model to convert it from a rocket to a glider. This is usually accomplished by either changing the location of the model's CG or changing the model's physical characteristics, or a combination of both.

While the Estes–Schutz B/G reduced its weight by motor ejection (now no longer permitted) and thereby reducing its wing loading, the ejection of the motor casing released a complex mechanical mechanism that permitted control surface—elevons—on the trailing edges of the wings to be deflected upward. From Figure 13-14, you can see that this produced a pitch-up force to oppose the pitch-down force of the wing lift.

Nearly all rear-motored B/G models (there are few rear-motored RGs) use the ejection of a propulsion pod or module to release control surfaces. However, these pods or modules must descend with a streamer or parachute attached because the NAR competition rules (and common sense) don't allow the ejection of an empty motor casing from a model rocket. (Not all model rocket motor casings are small, not all are made from paper, and all *are* difficult to see when falling free.)

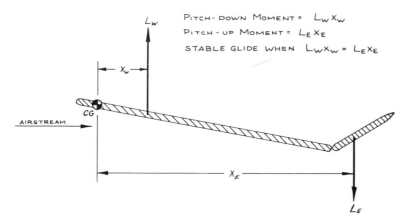

Figure 13-14 Use of an elevon control surface on a flying wing to provide stability. Wing alone gives pitch-down moment; elevon gives pitch-up moment. When the two pitching motions are equal, the wing will glide.

Rear-motored B/G models usually have delta-wing (and its derivative, the flying-wing), canard (horizontal stabilizer in front), or variable-geometry (swing-wing or scissors-wing) configurations. However, the simplest of all B/ G models to design and build—but not to trim and fly—is the front-motored conventional configuration rocket-boosted glider pioneered by Larry Renger similar to the one shown in Figure 13-15.

With movable-surface glide-recovery models, we saw how the pitch-down lift force of the wing with a forward CG was counterbalanced by the lift force generated by control surfaces to maintain a stable glide. But front-motored pop-podded B/G models and front-motored, CG-shifting RGs usually don't change any aerodynamic surfaces when the motor ejection charge activates. Instead, the CG of the model is shifted, and with the B/G model the wing loading is greatly reduced by the jettisoning of the power pod—or the separation of the booster rocket in the case of parasite B/G types. Thus, in CG-shift models, no new lifting forces are created at the transition from boost to glide, but the *relationships* between the lifting forces and the CG are changed by changing the CG location.

Basically, we change the pivot point of our aerial teeter-totter.

Figure 13-16 is a very simplified side view representation of a very simple CG-shift glider with all elements not germane to this discussion not shown (but you can imagine that the vertical stab and other elements are there if it will make you feel better). The wing is located on the forward portion of the model and a horizontal stabilizer is located at the aft end of the model.

During boost phase, the CG is located at or in front of the leading edge of the wing. The model therefore acts like a fin-stabilized model rocket because

Figure 13-15 A typical front-motored conventional configuration B/G, the author's original contest-winning Unicorn design with Vee-tail and pop pod.

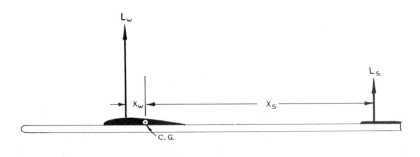

Figure 13-16 Representation of a CG-shift boost-glider in gliding condition with the CG shifted to 50% of the wing chord. The wing now has a pitch-up moment that is balanced by the pitch-down moment of the aft-mounted horizontal stabilizer.

the CPs of all surfaces are behind the model's CG. If, at transition, the CG was not shifted backwards, the model would continue to behave like a fin-stabilized model rocket and fly its parabolic trajectory right back into the ground.

But if the motor ejection charge jettisons a front-mounted pop pod containing the expended motor casing (B/G model) or moves the power pod rearwards (RG model) at transition, a totally different aerodynamic situation is created.

Let's suppose that the CG is shifted aft to a location which is 50 percent of the wing chord as shown in Figure 13-16.

The CG is now *behind* the CP of the wing and *ahead* of the CP of the horizontal stab.

If the wing were up there all alone, the wing lift concentrated at the wing CP would cause the wing to pitch up. You can create this situation by taking a piece of 1/16-inch sheet balsa or heavy cardboard about 3 inches wide and about 12 inches long; try to make it glide by itself. You'll find that its leading edge will pitch up, and the whole sheet will descend to the floor in a rotational flip, rotating about the spanwise axis at 50 percent of the chord.

To counteract this wing pitch-up force, a counteracting pitch-down force must be created to keep the model in level flight. This is done by adding the horizontal stab at the rear end of the fuselage. The horizontal stab is really just another little wing. It also produces lift, but its CP is a long distance behind the model's CG. This lifting force of the horizontal stab produces the necessary pitch-down force.

Although the lift force of the wing is much greater than that of the stab because of the greater area of the wing, the lifting force of the wing is closer to the model's CG. Therefore, the lesser pitch-down force produced by the smaller stab balances the wing's pitch-up lifting force. It's like balancing a 50-pound boy with a 250-pound middle linebacker on a teeter-totter. You can do it if you put the 250-pound linebacker closer to the pivot point than the 50-pound boy. This is because their *moments* will be equal.

What is a *moment*? Simply the product of the force and the distance through which it acts. It is weight times the distance from the balance point, or the force times the distance from the CG.

As you can see in Figure 13-16, the pitch-down moment of the stab (a small force acting over a long distance) is equal to the pitch-up moment of the wing (a large force acting over a short distance).

If the model starts to dive, the wing builds up more lift force as a result of increased airspeed than the little stab, and the forces become unbalanced in the pitch-up direction. The model pitches up to its balanced glide angle.

If the model starts to climb, the wing loses lift faster than the stab, and the pitch-down moment of the stab forces the nose down.

Yes, it's possible to have the wing and the stab exactly the same size.

This is called a *tandem wing* model. But the glide CG point must therefore be about halfway between the two tandem wings. The CG shift from boost to glide will have to be even greater with this tandem wing configuration, won't it?

The canard configuration can be looked at as a conventional model with a very small wing up front and a very large horizontal stab in back. The horizontal stab at the rear then contributes most of the lifting force of the total model, and it becomes the wing. Therefore, the CG shift for a canard configuration is even greater than that of a tandem-wing and much greater than a conventional configuration.

This is a highly simplified explanation of glide structures and aerodynamics. I had to digest a lot of data from technical reports and books before boiling it down to this point. But you don't have to be an aeronautical engineer to design a good glide-recovery model, nor do you have to be an expert to build and fly one. You do have to understand what we've discussed thus far, and you do have to be careful in your workmanship. And you should take your first steps in glide-recovery model rocketry with boost-gliders rather than the more complex rocket gliders. This is a very complex area of model rocketry.

BASIC DESIGN RULES FOR GLIDE-RECOVERY MODELS

To design a CG-shift glide-recovery model from scratch, one should have access to airfoil data such as in *The Theory of Wing Sections* by Abbott and von Doenhoff (*See* the Bibliography), which is a standard manual for aeronautical engineers. The method I used to make my calculations can be found in Frank Zaic's excellent and informative book *Circular Airflow and Model Aircraft*, also listed in the Bibliography.

However, you don't need to delve into these sources unless you want to, because I've worked out the following set of empirical design rules for front-motored CG-shift B/G or RG models. These rules aren't arbitrary. I didn't develop them by cut-and-try methods. I researched the literature, did a lot of design work in trade-offs and such, built models from the research results, tested the models in flight, and then came back and went around the circle again to refine the rules. They work for me, and they were put to the test by other model rocketeers, young and old, tyro and expert. They work. They'll work for you, too, if you follow them.

The basic Stine Design Rules are shown in Figure 13-17. This is a simplified three-view drawing of a hypothetical front-motored conventional configuration without its pop pod. It's not drawn to scale, but for maximum legibility.

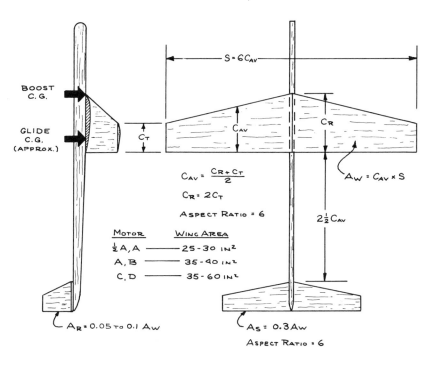

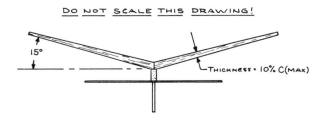

Figure 13-17 Stine's Basic B/G Design Rules.

Start the design with the basic dimension: the average wing chord. The wing span should be five to eight times the average chord dimension. This will give your glider wing an *aspect ratio* of 5 to 8, as defined in Figure 13-18. Although high aspect ratios of ten or more are used on high-performance model and full-size sailplanes, it's very difficult to build a high aspect ratio B/G or RG wing that will stay together during the high-acceleration, high-airspeed boost phase of flight. Only a few percent increase in performance and decrease in wing induced drag can be obtained with aspect ratios higher than eight.

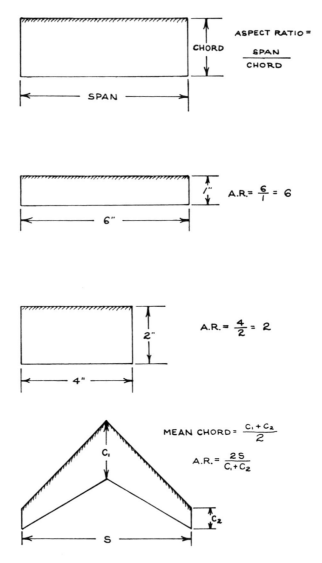

Figure 13-18 How to find the aspect ratio of a wing or fin.

For the best glide performance, a wing with a sweepback of 20 degrees or less—preferably none at all—should be used. High sweepback angles create high induced drag. *Moderate* sweepback angles up to 20 degrees may increase the model's yaw stability, but good yaw stability can be obtained by a properly sized vertical stab.

Although an elliptical wing shape such as that used on the legendary Supermarine Spitfire is the optimum aerodynamic wing shape for lowest induced drag and best stall characteristics, this optimum elliptical shape can be approximated within 2 percent using a straight taper wing as shown. The taper wing is similar to that used on an equally legendary airplane, the North American P-51 Mustang fighter plane. The benefits of an elliptical planform can be obtained with a taper wing where the tip chord is one-half of the root chord, as shown.

Although a glide-recovery model will fly with an unstreamlined, unshaped, unairfoiled slab of sheet balsa for a wing, it will fly better and be somewhat easier to trim if an airfoil is used. This is because an airfoil with a rounded leading edge and a tapered trailing edge will produce a higher lift-to-drag ratio than a flat plate. And it isn't really necessary to use the classical model airplane airfoil that's curved on top and flat on the bottom; a symmetrical airfoil such as used on fins will glide just as well (and perhaps better since its CP remains at or near the 25 percent chord point, whereas the CP of a flat-bottomed or *cambered* classical airfoil moves *forward* with increasing angle of attack, introducing another variable into an already complex system).

The thickness of the wing should be between 5 percent and 15 percent of the wing chord, and the maximum thickness of the airfoil should occur at 30 percent to 40 percent of the chord. If the wing is tapered in planform, it should also be tapered in thickness to preserve the same airfoil all along the wing.

Many model rocketeers and nearly all hand-launched glider experts believe that the thinner the wing airfoil, the better. This doesn't jibe with wing theory or with actual field data. The thicker the airfoil, the greater the lift force per unit area up to a thickness-to-chord ratio of 15 percent for most airfoil shapes. A thicker wing is a stronger wing, although a thicker wing will offer higher drag than a thin wing. A hand-launched glider has a propulsion phase of flight about four feet long—the distance through which your arm moves as you heave the glider. Any glide-recovery model rocket has a propulsion phase under rocket boost that's perhaps several hundred feet long. Although a "fat wing" B/G or RG may gain less altitude during boost because of the higher drag of its thick wing, this can often be more than offset by the improved glide characteristics and the higher strength of the thicker wing. One of the biggest problems with thin-wing B/Gs and RGs is their lack of strength in comparison to thick-winged models, which results in a higher risk of striptease during boost. Championship full-sized sailplanes have thick wings in comparison to classical hand-launched gliders of model aviation. So do full-sized airplanes. My manned boost-glider (a Piper Cherokee 140B) has a wing with 15 percent thickness at 40 percent of the chord, an aspect ratio of 5.71, and a glide ratio and therefore a lift/drag ratio of 10.32—which isn't the best in the world, but comparable to most hand-launched gliders and glide-recovery model rockets. (As a matter of fact, I've checked out a lot of glide-recovery dynamic stability hypotheses using my Cherokee because it offers the advantage of having the observer aboard rather than watching from the ground. It's also controllable.)

The horizontal stab of a glide-recovery model should have an area of 30 percent of the wing area, plus or minus 5 percent. It shouldn't have more than 2 to 4 degrees negative incidence or downward tilt with respect to the wing; you'll find 5 to 10 degrees of negative incidence in most inexpensive "chuck gliders" you can buy in a hobby shop, but this negative incidence was put in to insure maximum pitch stability when flown by anybody. A zero-zero glider—one with the wing and the stab at the same incidence as shown in all the examples illustrated thus far—will glide fine. However, thin-wing zero–zero designs may have a tendency to get into a dive and refuse to pull out, particularly if the transition from boost to glide occurs with the model in a nose-down attitude. A very slight 2- to 4-degree negative incidence will prevent this from happening, but also requires a higher wing loading and a more forward CG location during boost to keep it from looping under power. A thicker wing is a better solution because this creates more wing downwash on the stab. Important: for zero–zero models, put the stab behind and in the wake of the wing to take advantage of the fact that a lifting wing deflects the airstream downward behind it, creating what is known as down-wash. The wing downwash on the stab makes the stab think its flying at a negative angle of attack, and greater wing downwash eliminates the need for negative stab incidence.

The other design characteristics and dimensional relationships of the Stine Design Rules are exactly as shown in Figure 13-17. They're somewhat forgiving because you can change them 10 percent or so either way without getting into too much trouble with the design.

Note that the single vertical stab is mounted *underneath* the horizontal stab. Not only does this keep the vertical stab completely out of line of the motor exhaust jet from a front-mounted power pod, but its location under-neath aids the glider in rolling into horizontal flight at the top of its climb during transition. A top-mounted vertical stab not only operates in the wake of the horizontal stab at or near the stall point, thereby reducing yaw stability because a properly trimmed glider operates *just above* the stall point, but also doesn't contribute to roll stability but works against it.

These Design Rules will also produce a very fine chuck glider for hand launching.

When completed, the basic glider should have a weight of about 0.025 ounce (0.71 gram) per square inch of wing area.

For best performance, a B/G or RG should be designed for the motor type it will use. For B/G models, the following basic sizes have been found to be successful:

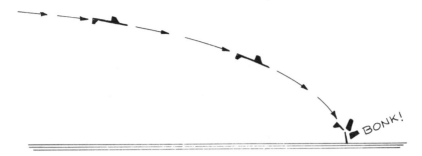

Figure 13-19 Nose-heavy gliding flight.

Motor Type	Wing area (Square Inches)
Type A	20-35
Type B	30-45
Type C	30-60

We haven't developed any sizing rules for larger B/G models yet. And there are no basic sizing rules for RG models at this time.

Boost-Glider Pop Pod Design

Although rocket glider models retain their motor casing and propulsion modules during glide, nearly all competition boost-glider models use what is known as a pop pod. A typical pop pod is shown in Figure 13-19.

A pop pod separates the entire weight of the propulsion unit from the model at transition, leaving only the basic glider portion of the model to glide. Anything that doesn't contribute to the glide characteristics of the model is thus jettisoned with the pop pod.

The pop pod shown in Figure 13-19 incorporates a streamer for recovery. Some of the larger pods for larger motors use parachutes. But I've learned from long and hard experience that a parachute will often cause the pod to get tangled with the glider. This happens often enough anyway, even with a streamer recovery pop pod, resulting in what is known as a "red baron" in which the whole model spins slowly to the ground with the streamer, pod, and glider tangled up and looking like the Red Baron himself shot the whole works out of the sky.

You want that pop pod to drop out of the glide path of the glider so they don't run into each other and get tangled up. A parachute tends to keep the pop pod up there in the glider's flight path.

Always use a motor hook to insure the retention of the motor in the pod because the basic action of a pop pod requires an energetic ejection of the streamer and nose; the reaction force of this forward-moving mass pushes the pop pod backwards on the model, unlatching the simple balsa hook on the plyon that mates with a notch in the forward fuselage of the glider.

Painting and Finishing Glide Recovery Models

Finishing a glide-recovery model is another area of great controversy where there's very little solid data to point in one direction or another.

Some modelers leave their gliders unpainted, sanding the glider smooth with No. 400 sandpaper. They maintain that the extra weight of a smooth finish detracts more from the glide performance than the aerodynamically rough finish.

Others use "glider dope" a mixture of 50 percent clear model airplane dope and 50 percent dope thinner, putting on several coats and sanding with No. 400 sandpaper between coats until a very smooth finish is obtained.

Still others use filler to remove all balsa grain, then paint the model with several costs of acrylic enamel with a spray gun or air-brush. The bright colors of these paint jobs often enable judges and timers to see the glider at a greater distance.

A popular finishing technique, especially in Eastern European countries and with advanced modelers flying with motors of Type C or larger, is to cover the balsa wing and stab surfaces with model airplane tissue, which is then doped to tighten and strengthen it. This can result in a very strong wing or stab.

The biggest problem faced by a modeler in finishing the relatively fragile parts of a glide recovery model is the fact that model airplane dope shrinks as it dries. This can cause large, thin surfaces on wings and stabs to warp. Therefore, some modelers add a few drops of castor oil to a bottle of dope to destroy this shrinking characteristic. *But it doesn't work with all types of dope,* so run a few tests first.

Trimming

A glide-recovery model must be balanced or "trimmed" for glide before it's flown. There are very few glide-recovery model designs than can't be flown without preflight glide trimming; delta-wing types often must be trimmed by means of a series of actual flights with low-powered motors.

Trimming a B/G is easy. Use the glider alone without the pop pod mounted or without the propulsion module installed. In other words, the gliding portion must be trimmed in its gliding configuration.

A grass-covered field is best for trimming because it keeps the glider from getting dinged if it dives into the ground. Naturally, if there's only one rock in the field, the glider will be drawn to it like a magnet and hit it. And, naturally, the glider will always break in the worst possible place.

The first trim flights should be gentle. Don't heave the glider! Grasp it behind the wings or in a place near the CG where you can get a good grip on it. Toss it gently away from you in a horizontal attitude with an overhand motion and release it into a glide path just *slightly* below the horizontal. Make several tosses to confirm what the glider does. You may not have tossed it correctly the first time.

If the glider dives as shown in Figure 13-20, it's nose-heavy. You must therefore remove some weight from the nose or add some to the tail, or both.

If the glider stalls as shown in Figure 13-21, it's tail-heavy, and you'll have to add some weight to the nose. This is the most common pretrimmed condition of any glider: tail heaviness. And that's why I always hollow out a compartment in the nose of my gliders so that I can put weight inside the fuselage where it won't create a lot of drag.

The best thing to use for trim weight is a little glob of plasticene modeling clay that can usually be purchased in the crafts section of the hobby or toy store. Put a hunk in your range box. To use it, pinch a little onto the nose or tail of the glider, as the case warrants. Or put it into the nose compartment as I do. Add a little bit of clay and make another trimming flight. Add or take away a little clay until the glider sails away from you in a slowly descending glide path. A normal B/G design should land 15 to 30 feet away from you, depending on how tall you are and upon the design of the model.

Now add just a teeny pinch of clay to the *left* wing tip to cause the glider to turn slowly to the left as it glides. Reason: once the glider gets a couple hundred feet in the air after boost phase, you don't want it to sail off on a straight cross-country mission. A straight-flying glider will nearly always turn its tail into the wind and take off for the next state, flying downwind much faster than you can possibly run. A motorcycle or booney-bike is very helpful in this situation.

What you want is a glider that turns slowly over the launch area like a hawk circling its prey.

Why a left turn?

On warm and hot days, rising bubbles and columns of hot air called thermals are generated in most open fields because a hot spot on the ground heats the air directly above it. The heated air is less dense than the cooler air around it, and it starts to rise. It may rise either as a bubble or as a column of air. A thermal is a small low-pressure cell like a miniature storm system, except it isn't big enough to have clouds in it. Thermal bubbles are like

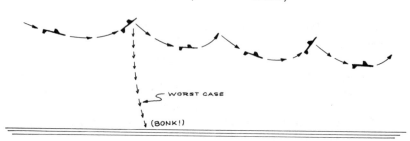

WORST CASE

(BONK!)

Figure 13-20 Stalling, tail-heavy gliding flight.

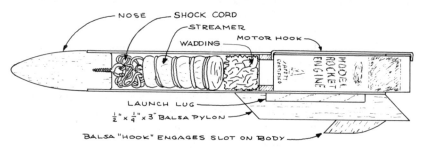

NOSE   SHOCK CORD
STREAMER
WADDING   MOTOR HOOK
MODEL ROCKET ENGINE
SAFETY CERTIFIED
LAUNCH LUG
½" x ¼" x 3" BALSA PYLON
BALSA "HOOK" ENGAGES SLOT ON BODY

Figure 13-21 Cross section of a typical pop pod.

doughnuts, and thermal columns are weak dust devils and very weak tornado funnels. When viewed from below in the northern hemisphere of earth, they circle counterclockwise. A glider that turns right will stick its nose directly into the thermal airstream and stay in the thermal, rising with it. A left-turning glider will fly through or out of a thermal, and you'll be able to get it back. I've seen B/Gs lost during trim flights, disappearing going up after being caught in a thermal.

Once having trimmed for a gentle, left-turning glide, the model's ready for a heave test. Haul off and heave the glider straight up as hard as you can. Good; the wings stayed on. The glider should climb straight up, losing speed, roll to horizontal attitude, and glide in a left circle about 25 to 50 feet in diameter. If it doesn't, try again. If it still won't perform right, remove just a *little bit* of nose weight and make it *slightly* tail-heavy.

Don't worry if your glider won't pass a heave test; some designs won't, no matter how hard you try. If, after persistent attempts to get a successful heave test, you cannot get the blasted thing to roll out and glide, go ahead and try it in a powered flight anyway.

Now balance the model for boost flight. Add the pop pod or propulsion module with an unused motor installed. The model should balance at or ahead of the leading edge of the wing, or at the point the kit manufacturer specifies in the instructions.

(You threw the instructions away after you built the model? You didn't put the instructions in your notebook? Didn't you have the presence of mind to read the instructions thoroughly, to know that you might need them when trimming the model? Didn't you think that you could build another model just like the first from parts if you traced the planforms on a sheet of paper and kept the instructions to tell you how to build the thing again?)

If a B/G model doesn't balance at or ahead of the leading edge of the wing or at the point specified by the manufacturer, add or remove weight to or from the pop pod on the nose of the pod. It's impossible to do this with a rear-motored model using a propulsion module, which is one reason why rear-motored models aren't popular. If your rear-motored model is tail-heavy in boost configuration, you've got to take weight out of the propulsion module, not off the model because the model's already been trimmed for glide.

If you've got a B/G or RG that uses surface-change, i.e., control surfaces that move at transition, check to make sure that the surfaces are locked into proper position for boost flight. If it's a kit, check the instructions concerning this.

Trimming an RG model is similar to trimming a B/G. Trim for glide first with an expended motor casing installed and everything in glide configuration. An RG may glide faster than a similar B/G because of the extra weight it carries, so you may not be able to glide-test it in any other way than tossing it from a second-story window to get enough altitude for an observable glide.

Carefully check the boost phase balance of any RG. This is where most RG models get into trouble. The first launch of an RG should be a heads-up affair because RG's have been known to loop and attempt to part the hair of the modeler or bystanders.

Flying Glide Recovery Models

Now you're ready to launch.

Make sure the pop pod is free enough to come off at transition when the ejection of the nose and recovery package kicks it rearward. Make sure it won't come off during boost; dangle the model from the pop pod to check this point.

In other types of B/G and RG models, make sure that the transition mechanism, whatever it is, is set to work as it's supposed to. Follow instructions for kit models.

A front-motored model normally needs to be held up off the base of the launch pad and jet deflector. A clothespin or piece of tape wrapped loosely around the rod (you may want to get the tape off for the flight of another model later) will support the model properly. Give it all the rod you can by

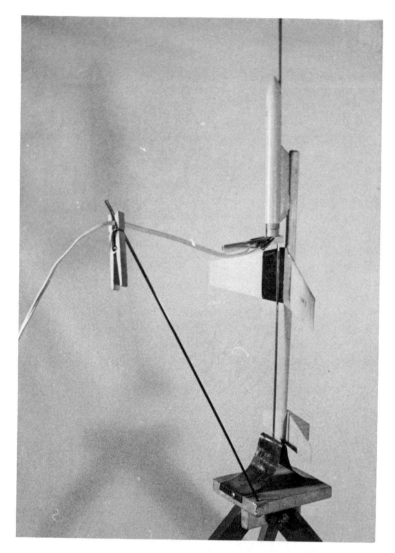

Figure 13-22 An umbilical mast installed on a launcher to support the ignition wires and prevent them from fouling the tail section of the B/G at lift-off.

making sure the tail of the model just barely clears the launch pad base or jet deflector.

Use an umbilical mast to hold the ignition wires and clips for a front-motored model as shown in Figure 13-22. An umbilical mast allows the leads to fall freely *away* from the model so they don't get caught on the model's tail section, for example. You can also tape the leads to the launch rod to keep them from falling into the model, but be sure that the model will rise up the rod clear of the wires and tape. It's very embarrassing to have the leads tangle with the model, especially in a contest where the rules say that such an aborted launch counts as an official flight.

If there's a wind blowing, it may push the model around on the launcher. Try to steady it with a clothespin or a piece of tape to prevent it from being blown or to keep the wind from blowing a B/G glider off its pop pod. But make sure the model will go up the rod upon ignition.

It's always a good idea to have somebody to help you on recovery, especially with a B/G. Your partner should go after the B/G pop pod or propulsion module while you chase the glider. Don't ever count on range people or bystanders to watch for your pod or booster. Basically, there are two types of glide-recovery model rocketeers: those with a large collection of pop pods with no gliders to go on them, and those with a collection of gliders whose pop pods were never found.

Yes, it's possible to design, build, and fly glide-recovery model rockets using the time-honored model airplane technology of dethermalizers. A dethermalizer is an on-board timer device which, after a given number of minutes, does something to the model to destroy its good glide characteristics and make it sink more rapidly to the ground. But this is an advanced area of glide recovery.

RADIO-CONTROLLED GLIDE RECOVERY MODELS

One of the first things a person asks about model rockets is: How about installing radio control? It's possible, but it seems to be practical only for glide-recovery models.

During powered or boost flight, things happen very fast. It takes a very good pilot to exercise effective control over a normal model rocket from the ground during boost. With a glide-recovery model, however, competitors on the national and international levels are now using radio control during the entire flight of glide-recovery models.

Radio-controlled B/G and RG models represent not only a very expensive area of model rocketry, but what is probably the most advanced. Modern solid-state electronics have greatly reduced the size and weight of radio control equipment. Radio-controlled B/G and RG models powered by Type D and larger motors are routinely being flown, but that whole area is beyond the scope of this book.

As of the 1980 World Space Modelling Championships, the big international competition is now between radio-controlled B/G models where the

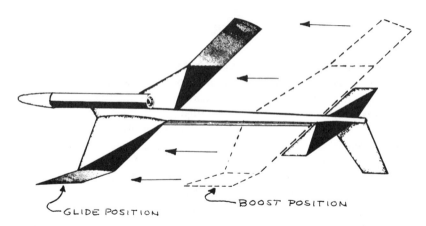

GLIDE POSITION

BOOST POSITION

Figure 13-23 Howard Kuhn's Buzzard rocket glider (RG). Wing is held back at rear by thread during boost phase. Motor ejection burns thread, and rubber band pulls wings forward on fuselage for glide phase. Nothing is ejected from the model. This successful RG is a Competition Model Rockets kit.

Figure 13-24 Guppy Youngren (right) was America's first gold medalist in world B/G competition. Here, Guppy and Randy Ringner prepare Guppy's Dark Star IV radio-controlled B/G at the World Championships in Yambol, Bulgaria. (Bob Parks)

Figure 13-25 Guppy Youngren's Dark Star IV radio-controlled B/G climbs out for its gold medal flight at Yambol, Bulgaria. (Bob Parks)

Figure 13-26 Guppy Youngren controls the glide of his Dark Star IV B/G during its gold medal flight at the Yambol, Bulgaria World Championships. (Bob Parks)

United States is clearly ahead and large flex-wing B/G models that have been brought to a high level of performance and reliability by the Bulgarians. But by the time you read this, the situation in the fast-moving field of glide-recovery model rocketry will probably have changed because this part of model rocketry is right at the leading edge of the hobby right now.

14

Payloads

Most model rockets are built strictly for sport flying or competition. They carry no payloads other than their own airframes. But one of the reasons for the existence of present-day full-sized rocket-powered vehicles is their ability to lift payloads to very high altitudes in very little time. Dr. Robert H. Goddard's original 1919 treatise on rockets written for the Smithsonian Institution was entitled, "A Method of Reaching Extreme Altitudes"; Goddard saw the rocket first as a means of carrying scientific instruments to altitudes that couldn't be reached by airplanes or balloons and only secondly as a means of space travel. In times past and even today, rocket vehicles carry payloads such as explosive warheads, signaling devices such as flares, rescue components such as ropes and lines, and scientific instruments. The most important payload carried by rockets is people.

In model rocketry, we don't work with explosive warheads. This is forbidden by the safety codes and by all the rules and regulations regarding nonprofessional rocket activities. Explosive warheads are very dangerous, and the direct handling of explosives is not part of the hobby/sport of model rocketry. It takes a great deal of highly specialized training to handle these materials with any degree of safety. In addition, one must be completely familiar with fusing and arming procedures, something one learns only in the military services after many years of training and experience. It's all far too hazardous for the average person to handle. If you're fascinated by explosives and things that go bang, you're reading the wrong book; join the armed services and get the proper training as an explosives and demolition expert there. Don't use model rocketry as your training ground.

In model rocketry, we fly for fun and knowledge, not to conduct a small war.

Payload-carrying model rockets are the province of experienced, advanced model rocketeers, not beginners. Many things have been done with payload-carrying model rockets, including pollution patrol, smog control studies, and investigations relating to ecological factors because model rockets are a fast and inexpensive way to lift small cameras or temperature sensors to several thousand feet. Many things can be done by payload-carrying model rockets that haven't been done yet, so it's a wide open field for careful research, creative development, and extensive testing.

Model rockets aren't usually designed to carry any old kind of payload that happens to come along, although model rockets can and often are adapted to carry a wide variety of different payloads. Usually a model rocket is designed around its intended payload with the designer keeping in mind the size, shape, and weight of the payload plus any environmental factors that will affect the payload—acceleration, shock, vibration, heat, etc.

In nearly all cases a model rocket's payload is carried in the nose section of the model. It may be housed inside a hollow nose or placed behind the nose in a cylindrical compartment that's structurally part of the nose assembly which comes off the model at ejection.

By positioning the payload up near the nose, a better CG–CP relationship can be obtained due to the payload weight. This can be important in many payload models because the additional payload weight usually causes lower lift-off accelerations and velocities, both of which require the model to have excellent stability characteristics.

There are a number of payloads that are commonly carried in model rockets. They can be grouped into the following general categories:

1. Passive, or dead load, payloads.
2. Optical payloads such as cameras.
3. Electronic payloads.
4. Active on-board payloads.
5. Biological payloads.
6. Special payloads.

Although there's some overlapping among these categories, we can discuss them individually because their airframe and propulsion requirements are often quite different. You'll be able to see for yourself where the categories overlap.

PASSIVE, OR DEAD LOAD, PAYLOADS

When the hobby of model rocketry began in 1957, the main emphasis was on propulsion and airframe technology because *everything* was new and we early enthusiasts had to begin with very simple models and very simple techniques. We didn't have a large selection of model rocket motor types to choose from; in late 1957 and early 1958, we had only Type A4 motors, and we got Type B4 motors in late 1958. Therefore, we didn't have the propulsion capability to lift large payloads. Small payloads were a real challenge because transistors were just beginning to become commercially available and integrated circuit chips hadn't even been heard of yet. Today's miniature cameras were also years in the future.

The NAR recognized the basic payload-carrying potential of model rockets and developed what was then called the "passive payload" competition category. This is still the payload category that's flown internationally under the rules of the Federation Aeronautique Internationale (FAI). It's based on the ability of a model rocket to carry one or more Standard Payloads to as high an altitude as possible with a given amount of total impulse.

The FAI Standard Payload is a small wafer of lead with a diameter of 19.05 millimeters (0.75 inch) weighing no less than 28.35 grams (1.0 ounce). This was also the original NAR Standard Payload and was selected in 1959 because all the available body tubes had an internal diameter of 3/4 inch. The FAI Standard Payload is the old NAR Standard Payload converted into the metric system.

The current NAR Standard Payload is a nonmetallic cylinder containing fine sand with a mass of no less than 21.0 grams and a cylindrical diameter of 19.0 millimeters, plus-or-minus 1.0 millimeters. This change from the FAI standard was made by the NAR Contest & Records Committee in 1979 because they believed that a contest payload of sand was less hazardous than the FAI lead payload.

But regardless of whether you're flying the FAI or the NAR competition payload, all the other rules are the same and, to a large extent, so are the basic principles of passive payload model design, construction, and operation.

The rules go on to require that no holes be drilled in the payload, that it not be altered in any way, that nothing be permanently attached to it, that the contestant be able to insert or remove the Standard Payload from the model at will, that the payload be totally enclosed within the airframe of the model, and the model be designed so that the payload cannot separate from the model in flight.

In essence, the model rocketeer is being put into the shoes of a real rocket design engineer who's told, "Here's the payload. It weighs this much and has these dimensions. You cannot change its shape. You cannot alter it in any way. You don't even have to know what it is. Just build a rocket vehicle to take it as high as possible."

These passive payload specifications aren't very difficult to meet. Nearly any good sporting model rocket can be converted into a passable payload model by the addition of a payload compartment or payload nose assembly. But it probably won't win many contests. The model rocket designer tries to achieve the optimum design for a model that will meet all the requirements of the competition.

Figure 14-2 shows J. Talley Guill's gold-medal-winning payload model from the First International Model Rocket Competition held in Dubnica-

Figure 14-1 Model rockets can be designed to fly payloads such as a fresh hen's egg. Jim Starks prepares his egg-carrying model for flight. (James E. Gazer)

nad-Vahom, Czechoslovakia, in May 1966. It's still a good design for the FAI Class S2A Single Payload category.

The cutaway of a typical FAI Class S2A Single Payload model is shown in Figure 14-3. Note that standard off-the-shelf catalog-available parts are used throughout. In view of what we've learned in the preceding chapters about aerodynamics and shapes, note the little extra touches that contribute to increased overall performance by means of drag reduction techniques. The model has been optimized for one purpose only: carrying the FAI Standard Payload to the maximum possible altitude. The nose is hollowed-out balsa, or a hollow vacuum-formed plastic nose is used. Every bit of weight that doesn't contribute to drag reduction, structural integrity, and payload retention has been eliminated or reduced.

To fly this model in the NAR Class B or Class C Payload Competition, you'd have to install a nose assembly of slightly different design to hold the

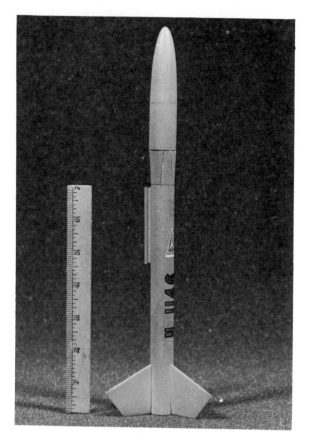

Figure 14-2 The United States' first international winner, Talley Guill's gold medal contest payload model which took first place at the First International Model Rocket Competition in Dubnica-nad-Vahom, Czechoslovakia in 1966.

NAR Standard Payload of sand in its plastic cylinder. This will just make the nose section a little bit bigger and longer. Competition Model Rockets sells an NAR Standard Payload, including the sand.

The nose plug fits into the payload shroud very tightly. The best way to get this kind of fit is to wrap cellophane tape around the nose plug until it slides very tightly into the payload shroud. This is known as shimming. The nose plug's tight fit into the shroud prevents the payload from separating from the model.

I've built and flown a lot of competition payload models, and it's impossible for me to say, "This particular design is the very best payload model design possible." There are too many trade-offs. I *do* know that a short, squatty

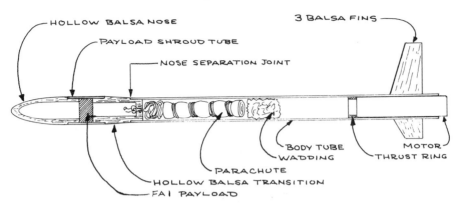

Figure 14-3 Cross section cutaway view of a typical high-performance competition model built to carry the FAI Standard Payload weight.

model will oscillate too much in flight, as we previously discussed. I also know that a payload model must have very efficient fins that will produce a high fin restoring force such as the shape shown in Figure 14-3. I also cannot tell you whether or not the fin shape shown in Figure 14-3, the long-span clipped delta, is better than a clipped delta fin with 60 degrees of leading-edge sweepback. This latter fin has less fin normal force but stalls at a higher angle of attack, whereas the long-span clipped delta has high fin normal force but stalls at lower angles of attack. Why the emphasis on angles of attack? Because the payload model with its heavy payload nose has a lot more angular momentum than an ordinary model rocket and therefore has the opportunity to swing to much higher angles of attack before the fins can produce enough restoring force. The competition payload model has what is known as a higher moment of inertia; once it starts to rotate in pitch–yaw, it develops considerably more rotational momentum than a sporting model.

Payload contest models are fun. Putting one ounce of payload up in the nose of a small model rocket *really* changes the way the model behaves. This is why it's been a competition event since the very beginning of model rocket competition.

OPTICAL PAYLOADS

One of the most interesting model rocket payloads is a camera. Many model rocketeers have worked very hard to build and fly camera models. The first camera-carrying model on record was built and flown by Lewis Dewart of Sunbury, Pennsylvania, in 1961. A small Japanese camera was simply strapped

Figure 14-4 The author prepares his contest payload model for its second place flight at the First Internationals in Dubnica-nad-Vahom, Czechoslovakia. (Otakar Saffek)

to the side of a model rocket; when the ejection charge popped the nose, it also pulled a string that released the shutter and permitted the camera to take a photo of the ground below—or the sky and clouds, depending upon the direction the model was pointed at that instant.

Estes Industries, Inc. brought out the first commercial model rocket camera, the original Cameroc, in 1965. The Cameroc permitted all model rocketeers to become in-flight photographers. The Cameroc lens pointed straight up through a transparent window in the tip of the round nose. Therefore, the model had to go over peak altitude and be pointed nose-down when the motor ejection charge popped the Cameroc nose off, releasing the shutter string at the same time the camera nose was popped off the model. The Cameroc took a single black-and-white photo 1.5 inches in diameter. Some

Figure 14-5 The Estes AstroCam® 110 camera model takes a single color photo on each flight using Kodacolor 110 film. (Estes Industries, Inc.)

modelers managed to fly color film in the old Cameroc. It wasn't exactly easy to load and unload the circular film in the old Cameroc, and you had to either develop the film yourself or send it back to Estes for processing because no commercial film processing outfit had the capability to handle the circular negatives.

In 1979, Estes Industries, Inc. introduced a greatly improved camera

Figure 14-6 Photo of Three Mile Island nuclear power plant, Middletown, Pennsylvania, taken with an Estes AstroCam® 110 by Scott Pearce of Westchester, Pennsylvania. (Estes Industries, Inc.)

rocket, the AstroCam 110 with a new Delta II launch vehicle tailored for the camera. The AstroCam 110 uses high-speed Kodak Kodacolor 110 color print film whose casettes are available everywhere and can be developed exactly the same as other Kodak Kodacolor 110 color print casette. The AstroCam 110 comes in kit form, so you have to build a camera; but it isn't difficult. If you're *this* far into model rocketry, you'll have no trouble at all.

It also occurred to many model rocketeers that a motion picture camera in a model rocket would produce a spectacular piece of footage as the ground fell away and the model climbed into the sky. The first in this area was a movie camera model built and flown by Paul Hans and Don Scott of Port Washington, New York, in 1962. This was a *big* model powered by a Type F motor because the smallest motion picture camera available at the time was the Bolsey B-8, a spring-wound 8-mm camera. It was heavy. Following months of preparation, including flights of test designs carrying dummy cameras, Hans and Scott committed their expensive Bolsey B-8 to flight. The lens looked out through a hole in the side of the nose section. The nose and body sections were recovered on separate silk parachutes.

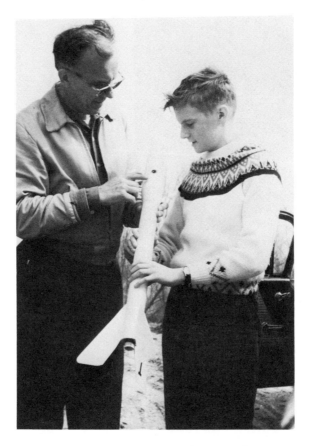

Figure 14-7 Charles and Paul Hans prepare the original movie camera model for its first flight.

On the first flight, everything worked perfectly. The model flew beautifully. Scott had to climb a tree to get the camera back. The color film was sent to the film processing lab—and was lost! The company replaced the film, but couldn't replace the flight footage. Hans and Scott tried again at the Fourth National Model Rocket Championships at the United States Air Force Academy in Colorado Springs, Colorado. This time the two model rocketeers personally took the film to a different processing lab with very explicit instructions.

That first model rocket in-flight motion picture film was indeed spectacular. The boys sold it to Time-Life, Inc. who never used it but left it to languish in their voluminous files where it remains today.

In 1969, Estes Industries, Inc. introduced the Cineroc®, a small, lightweight Super-8 model rocket motion picture camera developed by Mike

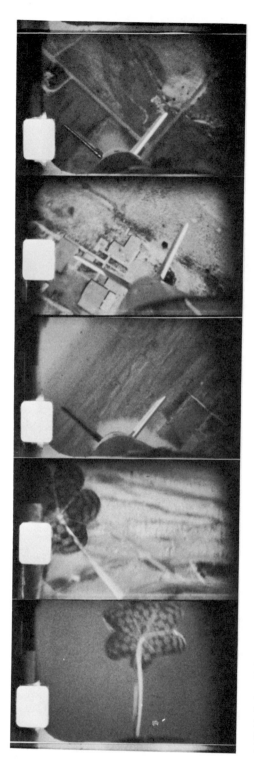

Figure 14-8 This is what it looks like from a model rocket in flight. Frames taken from an Estes Cineroc movie camera model, no longer available. The lens looks backwards along the body to the fins, shows lift-off, arc-over at apogee, and ejection of parachute. Modelers must now build their own cine camera models. (Estes Industries, Inc.)

Dorffler. The Cineroc was one of the most elegant model rocket products ever put on the market, and thousands of Cinerocs were flown by model rocketeers all over the world. Some incredible motion picture footage was made by the simple, battery-powered Cineroc camera. The Cineroc was available for about ten years. But the little Japanese electric motor that ran the Cineroc went out of production, and Estes stopped making and selling the camera. Rumors about the possible future availability of an improved, redesigned Cineroc-II persist, but there's been no announcement from Penrose, Colorado, as of this writing.

One can only hope that Estes or some other model rocket manufacturer will eventually put a model rocket motion picture camera back on the market. In the meantime, the gadgeteer rocketeers are back at work designing and building their own in-flight movie cameras again.

ELECTRONIC PAYLOADS

Real research rockets flown by NASA and other organizations engaged in rocket-powered experimentation and scientific research carry small radio transmitters to convert electrical signals from sensors—temperature and pressure pickups, for example—into radio signals that are sent to receiving stations on the ground. This gives people on the ground a moment-by-moment picture of what's happening up where the rocket is.

Model rocketeers have tried this, too. But it's no simple project, even today. One must have a thorough knowledge of electronics to get any meaningful information from a rocket-borne radio transmitter.

The first model rocket transmitter was designed and built by John S. Roe and Bill Robson of Colorado Springs, Colorado, in August 1960. About a year of work went into its design. The initial version was a simple transmitter whose schematic diagram is shown in Figure 14-10. It's only a single-transistor radio-frequency oscillator broadcasting on the 27-megahertz Citizen's Band. No information is sent by this simple transmitter. It merely emits a radio signal that indicates, "Here I am!' By using a directional antenna on the ground, you can find the transmitter and the model rocket after it's landed.

The circuit was then modified to produce the schematic shown in Figure 14-11 while the layout is shown in Figure 14-12. This unit was featured in the May 1962 issue of *Electronics Illustrated* magazine, which gives you some idea of how long the basic idea has been around. Roe added a pair of general purpose transistors wired as a free-running multivibrator whose frequency was changed by a change in the resistance of a sensor. A typical sensor would be a photocell looking out through a hole in the side of the rocket and reporting via a resistance change every time it saw the sun.

Figure 14-9 Bill Robson loads the first radio-carrying model rocket into a launch tower at the Second National Model Rocket Championships. Note long antenna wire coming from tip of nose.

Roe and Robson first flew their transmitter with such a photoconductive cell as a sensor. When the model rolled in flight, the photocell reported every time the sun shone on it. This changed the audio frequency impressed on the r–f carrier.

This is a very old design. Undoubtedly some electronically minded reader can come up with a modern version using integrated circuit technology.

Electronic payloads were a very active part of model rocketry about ten years ago. Estes Industries even had a Rocketronics Trans-Roc kit available. But the big problem at that time—and even today—is what to do with the transmitted information once you've received it on the ground. Some people have recorded it on portable tape casette machines which they've later fed

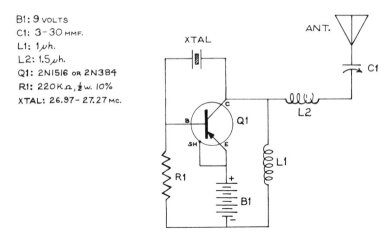

Figure 14-10 Schematic of Roe's simple model rocket radio transmitter.

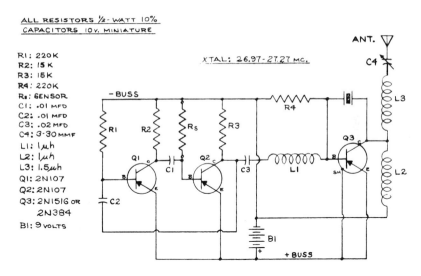

Figure 14-11 Schematic wiring diagram of John Roe's single-channel AM-FM model rocket radio transmitter.

into minicomputers as raw data; the minicomputer then displays or prints out the reduced data on temperature, roll rate, acceleration, etc., that the sensors on board the model had picked up.

The whole area of model rocket electronic payloads is, in 1981, a field ripe for reawakening, application of new electronics technology, and new

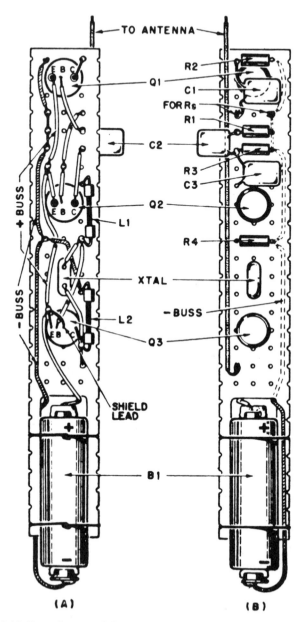

Figure 14-12 Parts layout of the single-channel AM-FM model rocket radio transmitter.

developments based upon the recent, ready availablity of minicomputers and microprocessors.

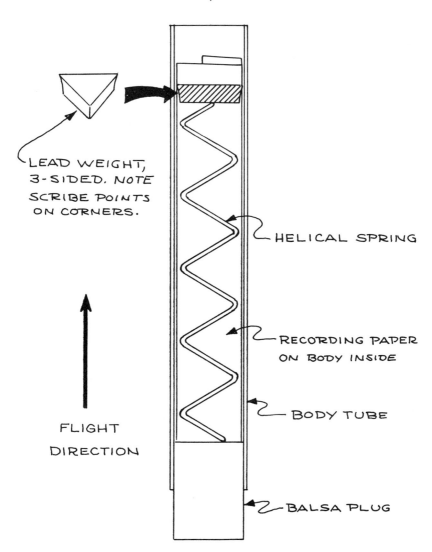

LEAD WEIGHT,
3-SIDED. NOTE
SCRIBE POINTS
ON CORNERS.

HELICAL SPRING

RECORDING PAPER
ON BODY INSIDE

BODY TUBE

BALSA PLUG

FLIGHT
DIRECTION

Figure 14-13 Cutaway sketch of Lindsay Audin's on-board accelerometer.

ACTIVE ON-BOARD PAYLOADS

The difference between this category of payload and the electronic payload is basically the place where the information or data is recorded for future use. The electronic payload transmits the data to the ground where it's recorded;

the on-board payload records the data in the model itself so that the information becomes available when you recover the model. To some extent, camera payloads fall into the latter category, although they have been discussed separately.

The primary active payload flown to date has been an on-board accelerometer that fits into the nose section of a model rocket. It's nothing more than a small triangular-shaped weight with a scribe point on it, a spring that supports this weight, and some scratch-sensitive chart paper that's wrapped around the inside of the payload compartment. The first of these simple on-board accelerometers for measuring the maximum acceleration of a model was built and perfected by Lindsay Audin of Hillside, New Jersey, in the early 1960s. It was similar to the one shown in Figure 14-13 which works as follows:

When the model accelerates at lift-off and during powered flight, the weight compresses the spring by a known amount that you've previously determined by ground calibration before the flight. The scribe on the side of the weight makes a mark on the sensitive paper inside the payload section. After recovery, you remove and unroll the paper. You can then see how far the weight compressed the spring because of the model's acceleration. Using the preflight calibration data, you can determine the maximum acceleration attained by the model.

There has been one known attempt to use modern microprocessor technology to electronically record data in a memory chip in a microcomputer aboard a model rocket. This was done by Charles Hall who reported on it at the 1980 Kent State Spacemodeling Convention at Kent State University. The on-board microcomputer put the sensor data into a memory chip so it could be read-out by another unit on the ground when the model landed and was brought back.

There are probably many other kinds of on-board active payloads that could now be built and flown using modern microprocessor and integrated circuit technology. There's a lot of room left here for advanced work and inventive ingenuity. The surface has hardly been scratched!

BIOLOGICAL PAYLOADS

Sooner or later a model rocketeer gets the bright idea that it'd be fun to fly a mouse in a model rocket.

Such a payload is known as a Live Biological Payload, or LBP for short.

There have been a great many mice killed in model rockets, proving only that a model rocket is a very expensive mousetrap. But it doesn't work as well as an ordinary mousetrap because you first must find a mouse and entice the little beast into the model rocket's payload compartment.

Unless you're conducting a valid scientific experiment under the direct supervision of a biology teacher, there is *nothing* you can learn by flying an LBP. As a matter of fact, you can subject a biological specimen to the same environmental stresses on the ground without a rocket flight at all.

Thus, there's very little reason to fly any sort of LBP except to inflate the ego of the model rocketeer. And there are many other things that you can fly that are more ego-satisfying.

The NAR and the Hobby Industry Association do not support the flying of live animals in model rockets, and an NAR member can lose membership privileges by doing so.

The American Society for the Prevention of Cruelty to Animals (ASPCA) has even stronger objections to model rocketeers flying live animals. The ASPCA has completely shut down all model rocket activity in several schools and clubs by *court order* because live animal flights were reported in the news media.

There's a better way, and it's more fun.

Special Payloads

In 1962, Captain David Barr of the United States Air Force Academy proposed a great idea that was immediately adopted by model rocketeers. Captain Barr felt that an excellent test of a model rocketeer's ability would be the flight and recovery of a *fresh* Grade A Large hen's egg without cracking the shell.

Egg lofting turned out to be a fantastically fun area of payload model rocketry. It was and still is quite a challenge.

A fresh Grade A Large hen's egg weighs an average of 2.7 ounces (76.5 grams) and has a minimum diameter of 1.75 inches (44.5 millimeters). Today's quality-controlled egg factories produce eggs with close-tolerance weights and dimensions. These mass-produced eggs also have very thin shells, which adds to the challenge of egg lofting.

Not only is an egg a very fragile payload, but if something does go wrong you have a scrambled egg and not a dead animal.

The first successful egg flight was made by Paul Hans and Don Scott in their movie camera rocket at NARAM-4 in 1962. Since then, thousands and thousands of model rocketeers have successfully flown eggs, and we've learned a lot of tricks about doing it.

First of all, an egg is very strong in its longest dimension and very weak across its minimum diameter. However, if it's completely and solidly cushioned all around, it will withstand a terrific beating. Essentially, put a strong shell all around the weak egg shell, an external structure that won't deform

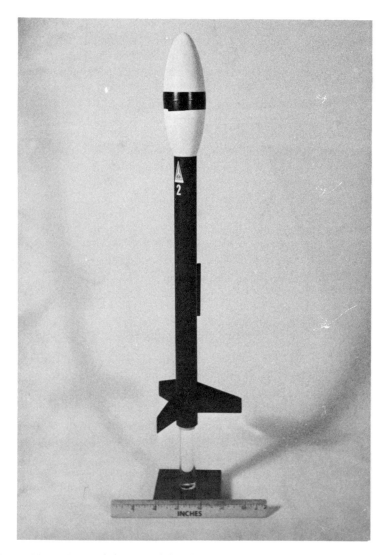

Figure 14-14 An egg-lofting model rocket using the Competition Model Rockets plastic egg capsule.

and crack the egg shell and that will *completely* cushion the egg to spread the impact stress.

Today's model rocket egg crates, often referred to as hen grenades, usually use a low-drag, vacuum-formed plastic egg capsule made by Competition Model Rockets (CMR). This capsule does such an outstanding job of protecting an egg that I've had my CMR egg capsule accidentally come loose from its parachute, fall freely for 1,000 feet out of the sky, land on packed soil, and not even crack the egg. To some extent, things like this take some of the fun out of egg flying. Still, with great regularity someone makes a terrific boo-boo resulting in a nose-down prang with a fresh egg that sends white and yellow goo flying in all directions. Some clubs keep a frying pan handy.

Within a year of Captain Barr's original suggestion, model rocketeers became so proficient at egg lofting that the NAR established an egg lofting payload competition category in which the contestant must have a model that will fly a competition provided fresh egg to as high an altitude as possible with restricted total impulse—and get the egg back to the judges unbroken, of course, or face the humiliation of disqualification and the agony of defeat.

Then model rocketeers got so good at it that they were regularly flying *two* eggs at once in one model. And at one contest, I saw the same egg flown *19 times* by 19 different modelers, which certainly must stand as some sort of unofficial record. (Do you know any egg that's flown 19 times in a rocket on the same day?)

Dual Egg Lofting was considered by NAR to be a bit "hairy," so Egg Loft Duration was set up as a category instead. This prevented model rocketeers from using the most common technique of egg flying: bringing the egg capsule down on a separate parachute from the rest of the model. With Egg Loft Duration, the whole model must come down tied together like an ordinary model rocket.

There's always great excitement in egg lofting, even when things don't work out exactly right. It certainly separates the good model rocketeers from the bragging balsa butchers. If you goof, you have a hilarious mess on your hands. If you succeed, you can cook your payload for supper.

SUMMATION

What else can be flown as a payload in a model rocket? Just about anything that weighs 6 ounces (170 grams) or less and will fit into any of the large variety of body tubes now available. There's not much to be gained by flying miniature kitchen sinks and other goofy payloads. But many things could be flown in model rockets that haven't yet been aloft under rocket power.

Payload-carrying model rockets present some interesting design and construction problems. They can be very complicated and very expensive, depending upon the kind of payload carried and the performance desired. In many cases, the model rocket is merely the inexpensive carrier for a very expensive payload. This is where the high reliability of model rocketry really pays off.

15

Scale Models

Since model rocketry is space rocketry in miniature, it's only natural that model rocketeers would want to make their models resemble the "big ones" as closely as possible. The construction and flight of exact miniature replicas of full-sized rocket vehicles and space vehicles—scale modeling—is model rocketry at its very best. It's the province of the true craftsman.

One of the first model rockets ever built was a scale model, a miniature version of the United States Navy's Pogo-Hi parachute target rocket. The builder was Chuck Moser, an engineer on the Pogo-Hi project at the Physical Sciences Laboratory of New Mexico State University, who completed and flew his scale model in March 1957.

Scale model rocketry combines craftsmanship with research. It reaches its pinnacle when a miniature replica, complete and correct down to the smallest details of marking and coloring, thunders off the launch pad for a straight flight and a perfect recovery. Months of research and work have probably gone into the model. Often such models are so beautiful that one hates to fly them. But having a beautiful scale model that's actually *flown* is much more satisfying than having one that sits forever earthbound on a shelf without ever being borne aloft under rocket power.

Scale models have been built with such fine workmanship and attention to detail that they qualify in all respects as museum scale models while also being capable of flight. Some model rocketeers even duplicated the launching facilities and equipment so that their scale model rockets take off from scale launch pads.

Many flying scale model rockets are in the collection of the National Air and Space Museum of the Smithsonian Institution. In some cases, they're the only models of a particular rocket vehicle that exist to help trace mankind's steps to the stars. As a result, they've become valuable additions to the national collection. The National Air and Space Museum has always been sympathetic to scale model rocketeers, some of whom have become the best astronautic historians alive today.

Scale model rocketry isn't something that can be done overnight. The creation of a scale model often takes weeks or months. It may take a long time simply to acquire the information that will ensure you really have a scale replica. And construction cannot be rushed. If you hurry, you're likely to put on that last frantic coat of paint on a perfectly built scale model only to have the paint blush or run, spoiling weeks of work. Don't start a scale model the night before a contest. Scale model rocketry takes weeks or months of planning, thought, careful workmanship, patience, attention to the smallest detail, and a lot of flight experience with nonscale sporting and contest models. Unlike participants in scale model aviation, scale model rocketeers are usually top-notch model fliers in their own right in addition to being scale modelers.

The rewards of scale modeling are great. There's an indescribable sense of pride and accomplishment when your model is placed on display after a flight. In contests, there are trophies and prizes to be won. In national and international meets there are always events for scale models because scale modeling brings out the best craftsmen, designers, researchers, and contest fliers.

Once you've become a scale modeler, all other model rockets seem to be just paper and balsa look-alikes.

The challenge of scale modeling never ends. When you think you've done everything, you'll always find another scale model beckoning to you. No modeler has ever built the perfect scale model. There are always improvements to be made, little flaws to be corrected on "the next one," or there are new techniques to try. New full-sized prototypes are always being developed, and these are fair game for the scale modeler. Some of the newer full-scale space vehicles such as the NASA Space Shuttle with its complex shape and tile-covered Orbiter pose real challenges to one's modeling abilities.

By participating in scale model rocketry, you're likely to become such an expert in one particular space rocket or vehicle that you could talk intelligently with the project manager himself. And you'll discover that you've become hooked on the pursuit of excellence and perfection.

IMPORTANT POINTS

There are several steps you should follow to build a good scale model. Occasionally, you'll discover an exception to these rules, but remember that these rules have been distilled from years of scale model work and two decades of my own personal experience in local, national, and international contests. Even if you don't wish to fly your scale model in contests, these points will serve you well and help you to achieve a perfect flying scale model of the prototype you're so fond of.

1. Select a good prototype to build a model of.
2. Get adequate scale data before you start to build.

Figure 15-1 The epitome of model rocketry is the construction and flight of a scale model that duplicates in miniature a full-sized rocket vehicle that flies. Here, a scale model Pershing tactical rocket lifts off. (Estes Industries, Inc.)

3. Using your data and the model rocket manufacturers' catalogs, select the correct size for your scale model, using as many commercial parts as possible; also select the type of model rocket motor that you'll use to propel your scale model.
4. Prepare accurate working drawings of the scale model and make accurate calculations of estimated CG, CP, weights, and flight performance.
5. Build a less-than-perfect flight test model first to ensure that your later highly detailed scale model will perform as you want it to. You'll also discover and eliminate some of the construction problems you're sure to run into.
6. Build two or three detailed scale models of the same prototype side-by-side at the same time.

7. Take your time, do your best work, and don't take shortcuts.
8. Flight-test your less-than-perfect test model, make any necessary changes or alterations, and always fly your detailed scale model before the contest.
9. Keep building progressively better scale models of the same prototype.

SELECTING A SCALE MODEL

The most critical and important step in scale modeling is the selection of the proper prototype to model. If you don't do this carefully, you may have considerable trouble in building the model or getting it to fly. Too many modelers have become discouraged about scale modeling because they made the wrong choice of a prototype. As a result, they ran into insurmountable difficulties with the project.

The first step in the selection process is an honest self-evaluation. You must honestly determine your own abilities. If you're not very good at construction, assembly, painting, or other workmanship factors, you should select a prototype whose scale model will not be beyond your capabilities in your weakest area. Incidentally, this also holds true for *all* model rocket activities, but it is "more so" in scale model rocketry. You shouldn't select a complex prototype to model if you don't think you can finish it; this is not copping out on a challenge, but simply being mature enough to recognize your own shortcomings and strengths in that stage of your development as a model rocketeer.

Nobody's going to ridicule you for starting out in a simple manner. In fact, it's smart to begin with a simple model that will result in a good, reliable, flyable scale bird. Don't let your personal enthusiasm for a certain full-sized rocket vehicle get the upper hand over common sense. You may be very eager to build a Saturn-Ib, a Titan-IIIC, or a Space Shuttle because you've seen one launched, photographed one in a museum, or otherwise have become taken with it. Or you may have been able to get all sorts of data on the old Nike–Hercules that's on display in the city park or the Atlas-D that's standing at the entrance of the nearby Air Force base. You may even have been able to get near a prototype to photograph it from all angles and measure it carefully. But do you really have the ability and experience to build a complex, highly detailed scale model on your first attempt?

The best model to start with is a simple, straightforward scaler that looks like the sporting model rockets you've been building. Your first scale model should be a single-staged vehicle with a cylindrical body, a simple ogive nose cone, and plenty of fin area. The color scheme should be simple, and the prototype shouldn't have a lot of detail on it.

Another important selection factor, which we'll discuss in greater detail later, is the availability of information on the prototype. Scale information on current military rockets is usually very difficult, if not impossible, to obtain because of military security precautions. This is particularly true of recent guided missiles. It may even be true of many historic rockets and guided missiles as well because the actual drawings and dimensions themselves may remain classified information even though there are hundreds of the rockets on public display all over the world. (It's also surprising what can be found in junkyards. I once spent $10 in a junkyard for a Nike rocket motor that I *knew* was classified. A friend of mine collected enough parts to build a *complete* Titan-II ICBM from dozens of junkyards from Boston to Baltimore.)

On the other hand, if you've chosen a very early rocket or space vehicle, you may have great difficulty in locating information because, believe it or not, pictures and drawings are often destroyed or thrown away to clean out the files and make room for the paperwork on the *next* project. Accurate information on many early rockets is difficult to find, even if the rocket was well-known. For example, I have literally traveled all over the United States and Europe, obtaining precise data on the old German A.4 (V-2) rocket. I've kicked myself from time to time because I was at White Sands when the United States Army was flying V-2 rockets there, and I've had my hands and tools inside V-2's that were launched a few hours afterward; thus, although I remember a lot of things, I never bothered at the time to collect scale data. So now I have to search it out, photographing one from a particular angle in the White Sands missile display, getting other data in the Deutchesmuseum in Munich, learning about another aspect in the Kensington Science Museum in London, and researching the library at the National Air and Space Museum. I know where *every* remaining German A.4 exists in the world. The only version I haven't seen is the Korolev V2B in the Kosmos Pavilion in Moscow, but I've got photos of it! I've worked years to accumulate all my scale information on the German A.4 and on more than a hundred other rockets. It isn't easy doing the historical research for these rare birds.

For those modelers who are starting in scale modeling, I've found it best to begin with a kit. There are several highly accurate scale model kits available from manufacturers as of 1981. If built to sound scale model standards, these kits will give you a good taste of scale models, will start you toward learning the competent construction and finishing required for scale models, and will help you produce a good-looking scale model capable of holding its own in most competitions.

There are several scale model kits available as of this writing. I hope there will be more available in the future. Unfortunately, some of the easier beginners' scale model kits have disappeared from the marketplace over the past few years. There are three kits currently on the market that can be made into good scale models, that are easy to build and fly, and that can get a modeler started in this fascinating field. The Estes No. 0862 A.S.P. can be

Figure 15-2 The real thing: a full-sized U.S. Army Pershing rocket lifts off. (U.S. Army)

modified into a good, simple scaler. The Centuri No. 5145 Nike Smoke can also be built as a good scaler with small modifications. And the Black Brant III® kits from Estes® (1293) and Canaroc® (54002) are outstanding.

Other good scale model kits on the market, some of which are complex, advanced, and/or expensive, are: Estes No. 1287 LTV Scout; Centuri No. 5131 Mercury-Redstone; Competition Model Rockets S1 D-Region Tomahawk; Estes No. 1269 Honest John; Centuri No. 5140 Saturn-Ib and No. 5142 Saturn-V; and the Centuri No. 5342 Jahawk AQM-37A.

If you want to build from scratch, try your hand on these prototypes which make excellent scale models: USN Viking #10 (on display at NASM), Sandia/NASA Nike–Tomahawk or NASA Nike–Cajun (on display at NASA Goddard Space Flight Center), USN Pogo-Hi (on display at White Sands Missile Range), or United States Army Jupiter-C (on display at NASM, the

Cape, and NASA Johnson Space Center). The biggest collections of prototypes for photographing and measuring are at the National Air and Space Museum, Washington, DC; U.S. Air Force Museum, Dayton, Ohio; the Cape Canaveral Air Force Base Museum and the NASA Visitor's Center, Cape Canaveral, Florida; NASA Johnson Space Center, Houston, Texas; Alabama Space and Rocket Center, Huntsville, Alabama; and White Sands Missile Range, New Mexico. Other smaller collections exist elsewhere.

When selecting a prototype, take into account these hints and tips concerning certain rocket vehicles. The USAF and USN Aerobee series should be avoided because of the booster rocket, which is attached below the vehicle by an open tubular framework, making it a difficult model to fly because, unless you've got some fantastic new staging system, all motors and recovery devices have to be in the booster, and the model must be flown as a single-stage bird. In fact, multistaged scale models shouldn't be attempted until you have a lot more experience because they're difficult and because there aren't that many good multistaged prototypes around. Air-launched missiles such as the United States Air Force Falcon series and the United States Navy Bullpup, Sparrow, and Sidewinder missiles should be avoided because they have guidance control fins up near the nose or were designed to be launched from airplanes flying at high speeds. Antiaircraft rockets and similar missiles typified by the United States Army Nike series, the United States Navy Tartar, Terrier, and Talos series, and the Soviet SAM missiles should be tackled only by experienced scale modelers. The Thor, Jupiter, Atlas, Titan, and Minuteman missiles are finless and therefore require the addition of clear plastic fins to make them aerodynamically stable scale model rockets, and they should be attempted only by experienced scale modelers.

Unless you're *really* very good indeed, stay away from the NASA Space Shuttle. It is *very* complex structurally, aerodynamically, and in appearance because of details such as the thermal protection tiles.

Even with these caveats, there are hundreds of rockets and space vehicles suitable for scale modeling, and new ones turn up all the time.

OBTAINING SCALE DATA

Once you've chosen a prototype to model—and choose two or three different ones just in case you run into difficulty at some later stage—the next step is to get information that will permit you to build a true scale replica and not just a semiscale bird that looks something like the real one.

Of course, you can build a "sport scale" model, which isn't such a bad idea for your first scale bird. A sport scale competition model must have all the substantiating data of the full competition scale bird, but the judges are not allowed to inspect the model from a distance any closer than one meter (3.28 feet or 39.37 inches).

You may have already done some research in the local library, which may have books with pictures and some very rough drawings. The NAR has a growing series of scale plans available to members (another enticement to join). A good source is a fellow model rocketeer who's collected photographs and drawings. As a matter of fact, a great deal of scale data swapping goes on in the NAR among scale modelers of all types.

However, to get really good data, you should attempt to go to the source— the manufacturer who made the prototype. If you don't know who made it, find out who uses it, such as NASA or the USAF. Send your first letter to the user unless you know who the prototype manufacturer is.

Your letters should be typed neatly on clean, white paper or written *legibly* in ink on good stationery. Don't scrawl a penciled note on a scrap of notebook paper. Include your *complete* return address, including your ZIP code. The results you get may depend upon the appearance, neatness, correct spelling, and correct grammar of your letter.

Don't concoct a fake "aerospace research center" letterhead or give yourself a fancy title. That doesn't help you at all and will, in fact, hurt your request. Everyone in the aerospace business knows everyone else and also knows what's going on in the business. A phony letterhead will be spotted immediately and will probably be chucked into the circular file along with the crank mail and nut letters.

Your letter should clearly state that you're a model rocketeer who wants to build a scale model. You should ask for specific data such as a color photograph and a dimensioned external drawing of the rocket. Be specific. Don't ask for all the information they've got on the particular prototype. You won't get it even if they're going to throw it away anyway. Remember, no aerospace company or government agency is going to turn themselves inside-out for you, taxpayer or youth science education notwithstanding. They'll probably send you whatever they're able to lay their hands on in a five-minute scan of the public affairs office and files, and the public affairs office is going to be the place where your letter ends up because nobody else really knows what to do with it or has the authority to send you what you've requested. Besides, aerospace companies and government agencies are always pretty busy and are always short of money. I'm not deriding aerospace companies or government agencies; some of them really will help you and have helped other model rocketeers in the past. I'm simply predicting their most likely behavior toward you so you won't get angry with them. Put yourself in their shoes, trying to answer dozens of letters like yours every day. Some companies and agencies are very good about sending data and photographs to model rocketeers, having anticipated your request and hundreds of others like it and gotten all the information together into "modelers packs." It may take several weeks to get an answer. If you don't get an answer to your letter in four weeks, write another letter, and please be polite.

The results of your initial request may be nothing more than a full-color brochure that's beautiful and expensive, but contains very little useful scale information. Write again, thanking them for the brochure and asking again for the specific information, explaining that you need it to build an exact scale model of *their* product or *their* launch vehicle that will eventually have *their* name on the side of it for everybody in town to see.

Above all, don't give up! Persistent, polite, intelligent model rocketeers have succeeded in getting valuable historical information declassified or rescued from eventual repose in the trash bin, thereby performing a highly commendable service to aerospace history.

Remember, it's *quality*, not *quantity*, of information that you want and need. Don't ask for the entire stack of factory drawings. There may be thousands of them amounting to several hundred pounds of paper, including details of every little nut and bolt that went into the prototype. You don't need all of that, and it's actually too much data. It's what is known as "wall to wall data" which is useless and necessitates hours or days of sorting to find what you really need.

Minimum Scale Data

Although you may not be entering your first scale models in contests, it's a good idea to follow competition rules from the very start when it comes to acquiring what is known as "minimum scale data." The consensus of the best scale modelers in the nation is represented in the NAR scale model rules in this regard, and almost anyone can obtain the NAR's minimum scale data. In fact, I really don't know how a scale model can be built with less than the required "minimum scale data."

Minimum scale data consists of:

1. Scale factor. More about this in a moment.
2. Overall length.
3. Diameter(s).
4. Nose length.
5. Fin length, width, and thickness, if prototype had fins.
6. Length of any transition pieces, if prototype had them.
7. Color pattern documented either in writing or by photographs.
8. One clear photograph, halftone reproduction, or photo-reproduction.
9. For all dimensional data listed above, both the actual prototype dimensions and the scaled dimensions used on the model, which must be presented either on a drawing or in tabular form.

Additional data are certainly desirable.

It's sometimes possible to measure, or "tape out," a prototype if you can get near it. If you do this, you should make careful notes as you go along, and then date the notes. A camera is an absolute necessity for recording details and shapes, and color film should be used for documenting color data. You might be surprised at the number of prototypes on display, just waiting to be taped out.

Scale data can sometimes be obtained from a professional or museum model, but don't count on it being accurate unless you can back it up from an independent source. Some models aren't accurate, even museum models. In spite of advertising claims, very few commercial, nonflying plastic model kits are accurate. Remember that the museum model builder or the plastic kit manufacturer has been faced with the same scale data acquisition problems as you. And because of deadlines, he's usually had to proceed using incomplete or even inaccurate data.

You should begin to collect and file scale data because its quantity soon begins to grow. Scale data becomes very valuable not only in building models but in trading with other scale buffs for data they have and that you need. Make a file folder or get a large manila envelope to hold all the information on a given prototype. Clip magazines and newspapers. Use copying machines to get copies of data you want. Scale data need not be original documents; good quality copies are useful and acceptable. Remember that you want and need quality data, not sheer quantity.

As you continue to collect scale data, you'll discover that you can't trust some of it. Data from different sources may not agree. You may have to do some careful research to determine which data source is correct. You'll undoubtedly run into The Problem of the Lost Inch where a mistake was once transferred from one drawing to another without anybody catching it; this may go on for years and years before someone, usually a scale buff, finally catches it. In some cases, I can spot exactly what data source was used by a modeler because of the mistakes in data that have shown up due to other data and additional research. I also know when somebody uses some of my early scale drawings by recognizing the mistakes that were in those drawings because I'd used the best available data at the time.

You should build a model of a particular prototype with a specific serial number, flight number, paint pattern, etc., unless the prototype was manufactured by the thousands on an assembly line with little or nothing except a stenciled serial number to differentiate individual vehicles. If the prototype wasn't mass-produced, make a model of a particular vehicle, because many variations can occur in individual versions of a prototype. For example, every one of the 14 United States Navy Viking rockets launched was different. Over 65 percent of the German V-2 rockets launched at White Sands had major external changes in their basic configuration, and every one of them had a

Figure 15-3 A photograph of the real I.Q.S.Y. Tomahawk as launched at NASA Wallops Station, Virginia. (NASA)

different paint pattern. The Freedom-7 Mercury Capsule was significantly different from the Liberty Bell-7, and if you build a Mercury–Redstone, you'd better have the proper Mercury capsule atop the properly marked and numbered Redstone booster. Most of the Saturn-I vehicles were different, and the paint pattern usually shown for the Saturn-V was not the one on the actual flight vehicles. If you get involved with a launch vehicle like the Delta where over a hundred of them have been launched, no two looking alike, you've got lots of research to do.

The best beginner's scale model I've ever found is the Thiokol-NASA I.Q.S.Y. Tomahawk whose dimensioned drawing, made from manufacturer's drawings, is shown in Figure 15-5. A photograph of the prototype Round 4 is shown in Figure 15-3, and a scale model of the prototype is shown in Figure 15-4. This has proven to be an outstanding beginner's model. A kit

Figure 15-4 Connie Stine hooks up the igniter on her scale model I.Q.S.Y. Tomahawk. She also built the scale launcher.

was once available, but no more. It's an easy scale model to build from scratch, and it flies beautifully.

DESIGNING A SCALE MODEL

When you've collected enough scale data to get started, you're ready to begin the design phase of your scale model, unless you're building from a kit. You should be completely familiar with the commercial model rocket parts that are available so you can use them wherever possible. You should know what model rocket motors are available and what their performance characteristics are so you can choose the proper one for your model. With all of this in mind, you can begin to size your scale model.

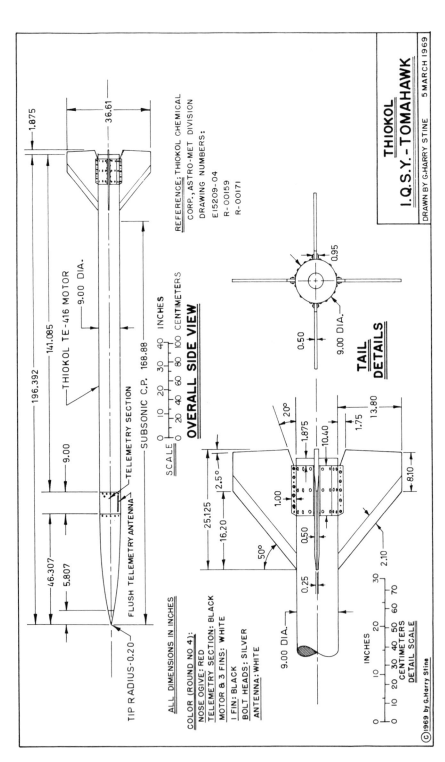

Figure 15-5 Scale drawing of the real I.Q.S.Y. Tomahawk. Use it to build your own scale model.

Sizing is a very important step. If you've chosen a prototype with lots of external detail and many complex shapes, you should build a big scale model of it. Little details are extremely difficult to make and attach to a small model. And on a small model details seem to get lost because the human eye cannot see them as well. On the other hand, if you've selected a rather simple prototype—basically, just a nose, body, and fins—your model will look better if it's smaller. In scale model competition, the overall appearance of a scale model is often just as important as its fidelity to scale and flight characteristics. All other things being equal, a large-scale model will usually gather more scale points than a small-scale model of the same prototype, even though the details on the small model required a magnifying glass to apply—and to see.

You may also have to size your scale bird on the basis of what you intend to do with it. If it's a pure competition bird, you don't have as many restrictions. But if you want a scale model that's also good for altitude performance,

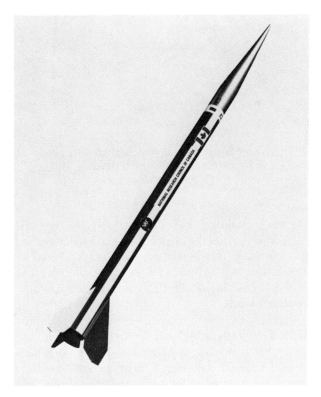

Figure 15-6 The Canadian Black Brant III sounding rocket is available as a scale model kit from both Estes Industries, Inc. and Canaroc Space Models. It's a good beginner's scale model. (Estes Industries, Inc.)

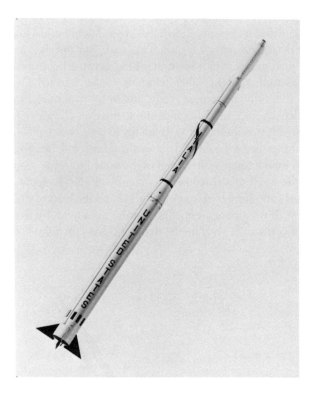

Figure 15-7 The NASA Scout launch vehicle is another good scale model kit available from Estes Industries, Inc. (Estes Industries, Inc.)

you'll look for lower total drag, high impulse-to-weight ratio, and a size and color scheme that will permit the model to be seen by altitude tracking crews.

Again, use as many commercial parts as you can. You'll have to make plenty of custom parts as it is, so don't make things difficult for yourself. Use commercial body tubes. Many other commercial parts can also be used or modified for scale models, too.

The biggest single problem with many beginners' scale models is their high weight. As a result, these big, underpowered scale birds make very hairy flights and some barely manage to stagger high enough into the air to crash. If your scale model doesn't make a good flight, it doesn't make any difference how accurate, detailed, or beautiful it is because the purpose of scale model rocketry is to build a scale replica that will fly.

So match your model to a motor before you start to design and build it, just as you'd match a motor to a sport model you'd designed.

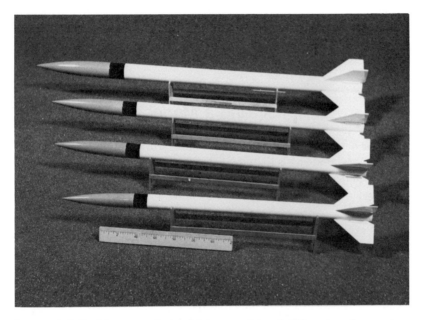

Figure 15-8 Building a scale model in a variety of different scales or sizes with body tubes of different diameters will help you get a good scale model that flies well. The author built these four I.Q.S.Y. Tomahawks in different scales to determine the best size.

There are no hard-and-fast rules for sizing a scale model because each case is different. Sometimes you may have to do a little experimenting by building two or more semiscale flight test models of the same prototype in differing scales to find out which size suits your needs—or which one flies best. I built the I.Q.S.Y. Tomahawk in four different scales with four different body tube sizes as shown in Figure 15-8, just to see which one flew best. They all flew equally well with the little one, of course, capable of the highest altitude. I've also built the ASP in three different scales for three different motor types, the Nike Smoke in four different scales, and the NASA Shotput in three different scales.

To design your scale model properly, you'll have to do a little engineering drawing or drafting. This isn't difficult, and you should learn how because it'll help you in your other model rocket activities. Today, drafting equipment is available at reasonable prices. You'll need a good drawing board, a T square, a couple of triangles, a good pencil with 2-H lead, an eraser, an engineer's triangular scale, a curve or two, a drawing compass, and a good protractor. Get a little practice first by designing and drawing a couple of sport models. You'll be surprised how easy it is to do model rocket drafting.

Having decided on the size of your scale model, you must now prepare a full-sized working drawing of the model so you can determine sizes and shapes of various parts. To do this, you must determine the scale of your model.

Scaling a Model

Scale refers to the relationship of a model's size to its full-sized prototype. The scale of the model is the ratio between a dimension on the model and the same dimension on the prototype. The most common method of determining scale is to compare the diameter of the prototype to the diameter of the model.

For example, if the diameter of the prototype is 31.0 inches, and you plan to use a body tube with an outside diameter of 0.976 inch, you divide 31.0 by 0.976 on your pocket calculator. The result is 31.762295. Round it off because the number 31.762 is close enough. The scale of your model will then be 1 to 31.762, written 1:31.762. It means that 1 inch on the model is equal to 31.762 inches on the prototype.

Using a piece of paper and your pocket calculator (a "constant divide" feature is great for this purpose), or using long division if you want to do it the hard, old-fashioned way, divide every dimension of the prototype by the scale factor—31.762 in our example. This will give you the dimensions of every part of the model.

Naturally, a 30-degree angle on the prototype is still a 30-degree angle on the model because an angle is not a dimension that changes with the size of the model. So don't divide angles by the scale factor, or you'll end up with a funny-looking bird indeed.

Using the dimensions you've calculated, draw a full-sized working plan of your scale model. A far easier and more accurate way is to take the prototype drawing to a blueprint or printing shop and have them make a photographic reduction or enlargement of the prototype plans to exactly the size of your model. Blueprint shops and printers work in terms of percentage enlargement or reduction; if the drawing of the prototype has a body diameter dimension *on the drawing* of 3.1 inches, for example—they've sent you a 1/10th sized drawing of the prototype—you'll want the drawing reduced by the factor 100– $[(0.976/3.1) \times 100]$ or 68.5 percent. Talk to them about it and explain what you want to do if you're confused; better to work with them to get it right than to pay for a photo reduction you can't use. This may cost a little bit more, but it save you having to do a lot of drawing and will be a lot more accurate.

From the scale working drawing you now have, you can determine the size of the parts. By tracing the fin outline onto a piece of stiff paper, you

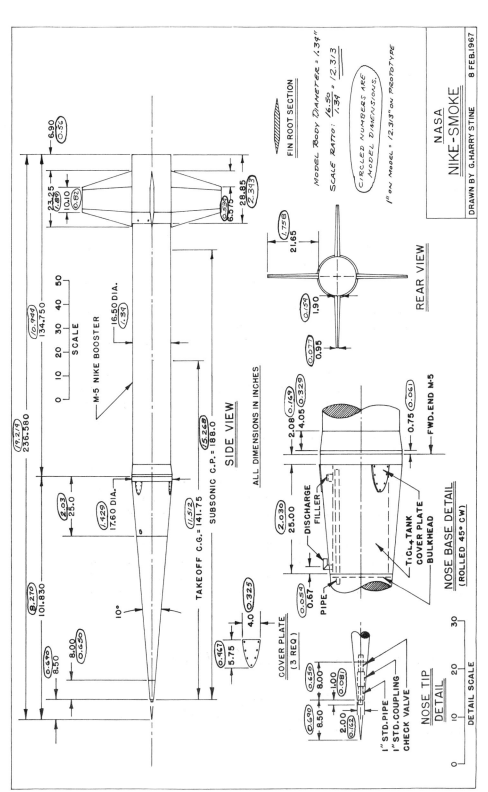

Figure 15-9 Scale model drawing of the Nike-Smoke rocket with both the prototype and the scale model dimensions scaled properly. Build it.

can make a template for cutting out the fins to the proper size and shape. You can also make paper templates for the nose shape, transition shape, boat tail shape, and other shapes of the model. These templates help assure that these parts are shaped correctly.

Don't start to build yet! Stay on the drawing board and the paper pad and resist the temptation to cut balsa until you've really got that model figured out.

Calculate the CP. It is not a good idea to use the cardboard cutout method or the swing test for a scale model because most scale models usually are a bit shy on fin area in the first place. If you use the cutout method, you'll probably end up with too much nose weight which will drastically reduce the flight performance. If you use the swing test, you've got the model built already and it may be too late to take corrective action to restore a proper CP–CG relationship. If you use the Barrowman Method or the STABCALC computer program, you'll be able to locate the CP very accurately ahead of time.

Make a trial run at estimating the CG using the weight of the various parts and their distance from a common point such as the nose tip. This is the same procedure as calculating the CP, except you use weight instead of area.

Calculate the flight performance using RASP-79E or an equivalent method, assuming several different values of Cd to bracket the altitude range in which your model will be flying. This will also help you make the proper choice of model rocket motor to be used. Sometimes you can obtain the actual Cd of the prototype which you can use in the calculations for your model because Cd doesn't change with size.

Why all this paperwork? So that you can discover ahead of time whether your scale model will be too heavy, underpowered, or require too much nose weight. You may also discover that if you built it, it wouldn't fly in a stable manner no matter what you did. You may even discover that it's impossible to build the model in the first place! In any case, time spent in preliminary paperwork pays off in scale model rocketry just as it does in full-scale rocketry. This is particularly true if problems show up in the design process. It doesn't take much time or money to redo calculations or change a drawing; it takes far more time and money to correct a mistake after you've started working on hardware.

BUILDING A SCALE MODEL

There aren't many special tips on building a good scale model if you've done everything correctly up to this point and are a careful model builder anyway. Just take care and do your best work.

Build a less-than-perfect semiscale flight test model first. Do a good job, but don't super-detail it. This is strictly a test bird. Build it and use it to check out your calculations, to see if it flies as anticipated, and to correct any little operating problems that you may discover by doing all this.

When you finally sit down to build the super-detailed scale model, build two of them at the same time. It doesn't take much more time and effort to make a pair, and it will give you a spare, JIC. Somebody might sit on your scale model two minutes before you get ready to fly it in that big regional contest.

You might also go back and bring your semiscale flight test model up to full-scale model standards, giving you three scale models of the same prototype.

IMPROVING YOUR SCALE MODEL

Most NAR and FAI champion scale modelers improve their scale models by building the same model over and over again in increasingly improved fashion. Every time you build a scale model, you learn little tricks and shortcuts you can use the next time you build the same one. You also improve your modeling technique and skill each time. Not only do the models get better, but you can build them quicker.

J. Talley Guill, many times United States Junior and Leader National Champion and FAI gold medalist, built models of the USAF-Convair MX-774 HiRoc for over five years, perfecting them to the point where they'd fly exactly as he wanted. His father, A.W. Guill, had the same thing going for the NASA Astrobee-1500. Charles Duelfer built models of the USAF GAR-11 Falcon with continual improvement. Howard Kuhn's Argo D-4 Javelins are works of art. Otakar Saffek of Czechoslovakia, 1972 and 1974 World Champion scale modeler, built over a dozen 1:100 scale Saturn-V models, improving them each time. His final achievement was perfect down to scale corrugations and took him 2,000 hours of work.

Once you get a good combination, stick with it and improve it. Do research on additional details. Lighten and strengthen the model. Strive for high flight reliability.

When you get tired of one scale model, put it away and try others, then come back to it later.

When you tire of building scale model rockets, try a scale model launch complex. These are undoubtedly the most impressive scale model systems in the world, often with remotely operated motorized moving parts such as

Figure 15-10 Scale model rocketry knows no country. The Polish scale model team at the First World Championships flew (left to right) a French Diamant satellite launcher, a U.S. Saturn Ib, and a USSR Soyuz. (Otakar Saffek)

launch rails that move up and down and turn in azimuth just like the prototypes. Others have service towers and gantries that move away under remote control and umbilical towers and cables that swing away just before ignition. Special effects experts have nothing on our advanced scale model rocketeers, some of whom have actually gone on to work on Hollywood's science-fiction special effects.

In scale modeling, there's literally no end to what you can do in miniature. It can give you a lifetime of enjoyment and challenge. When you're a good scale modeler, you're one of a very small and elite clan of model rocketeers.

We need more elite model rocketeers. Come on in! There's plenty of room for more people in scale model rocketry.

Figure 15-11 George Pantalos with his Thrust-Augmented Long-Tank Delta scale model at Vrsac, Yugoslavia. (Otakar Saffek)

Figure 15-12 The world's best scale model, a 1:50 model of the USSR Soyuz-33 built by Bulgarian gold medalist, Moritcz Mashiah, at the 1980 World Championships.

Figure 15-13 Lift-off of Mashiah's Soyuz-33 gold medal scale model.

16

Altitude Determination

Earlier we discussed methods of calculating the altitude performance of model rockets and learned that we can't accurately determine achieved altitude by calculation alone, modern minicomputers and programmable pocket calculators notwithstanding, because we can't control or accurately measure all the variables and parameters involved. Therefore, the only way to find out how high a model rocket design will go is to actually fly it and use one of several methods developed for altitude determination.

Because of the effects of aerodynamic drag, we can't determine the achieved altitude of a model rocket by timing its flight to apogee and assuming zero-drag characteristics. This should be patently obvious in light of our discussions and calculations of aerodynamic drag effects.

However, there's one altitude determination method which uses timing and which has proven itself to be reasonably accurate.

At the suggestion of Douglas J. Malewicki, then at Cessna Aircraft Company, and Larry Brown, then at Centuri Engineering Company, Bill Stine in 1974 decided to look into the possibility of determining altitude by timing the fall of a standard marker streamer with a fixed set of dimensions and a fixed weight. The theory behind this is that the marker streamer—a piece of 0.001-inch thick polyethylene film 1 inch wide and 12 inches long with a 3-gram nose weight taped at one end—would fall at a constant rate of speed. It would be packed into the model rocket atop the recovery device and ejected at or near apogee. The time required for the marker streamer to reach the ground from the moment of ejection would be equivalent to altitude. Naturally, this method would determine the altitude of the model at ejection, not necessarily peak altitude.

For sporting contests, every model would carry the standard marker streamer, and the fall time of each streamer would be scored. The model whose marker streamer took the longest time to reach the ground obviously ejected the marker at the highest altitude and therefore was the winner.

Some interesting facts were discovered when Bill Stine flew some experimental flights. Larry Brown had previously determined that the standard marker streamer would drop at a constant rate of 18 feet per second; he made a series of test drops from a known height on the Phoenix, Arizona, fire

Figure 16-1 Model rockets can be tracked in flight with very simple equipment to obtain a definite figure for the achieved altitude.

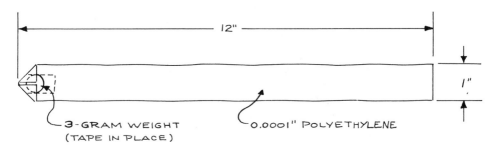

Figure 16-2 Sketch of the SAM (Standard Altitude Marker) Streamer developed and tested by Bill Stine.

department's training tower. Although Bill Stine recorded a spread in marker drop time of as much as 15 percent in identical models powered by motors from the same production lot, he came to the conclusion that this variation was a 15 percent difference in altitude caused by a variation in motor total impulse, and that the maximum altitude is a function of the square of the burnout velocity. Therefore, the 10 percent variation in total impulse per-

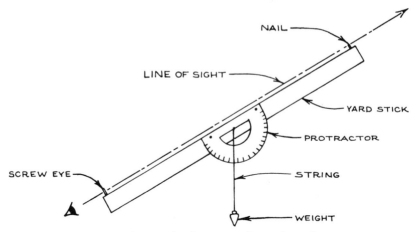

Figure 16-3 Diagram of a simple elevation-only tracking device.

mitted by NAR Contest Certification could produce a large variation in altitude.

It was not possible in these early tests of this altitude determination method to find out whether or not the fall time of the marker streamer provided good altitude data. Further work remains to be done, and it's a good science project for junior or senior high school students.

Fortunately, even though the marker streamer method is an excellent way to conduct an unofficial altitude contest with no equipment other than a stopwatch, we have other methods of determining with great accuracy the altitude achieved by model rockets.

When a big rocket flies at White Sands, Cape Canaveral, or Vandenberg, its flight performance and achieved altitude are determined by tracking the vehicle in flight with electronic systems such as radar or with optical devices like telescopes or laser rangers (lidar). Such tracking devices provide reasonably accurate data on the vehicle's position in space at any given instant.

In model rocketry, tracking is also used. However, electronic tracking of model rockets has not been used to date because of the very high cost of the equipment and the very small size of the models. Although model rockets were first tracked by radar at NASA's Wallops Flight Center in 1964, such electronic tracking appears at this time to be impractical because of the high costs and complexity of radar.

Therefore, optical tracking is almost universally used in model rocketry.

Optical tracking is just a fancy term for following the model in flight by eye and aiming some sort of measuring device at it. So far, the state of the art in laser tracking puts this method in the same league with radar, so optical tracking with "calibrated eyeballs" assisted by a device to measure angles is the method of current use.

Without a great deal of expense and knowledge, simple and reasonably accurate optical tracking equipment can be built by almost anyone with access to a junior high school shop or a well-equipped home workshop.

Optical tracking provides only one piece of information: a figure for maximum altitude achieved. This is an important datum, however, because it's useful in design evaluation, staging studies, and competition. Optical tracking combined with simple trigonometry ("simple" because you don't have to understand trig to make it work for you) has provided an easy, accurate means of obtaining this information since 1958.

There are three basic tracking methods. One of them is a "cheap and dirty" method that will give a general figure for achieved altitude using a single piece of simple equipment. The second is a highly refined and tested system of high accuracy using more complicated equipment; it's used world-wide for contests and record-setting purposes and is the FAI-approved system. The third system uses simpler tracking devices, has been tried and proven in the field, is as accurate as the FAI-approved system, but hasn't seen wide use because its simplicity hasn't been recognized by most people.

Modern minicomputers and programmable calculators have made it extremely easy to obtain data from all three systems.

The "cheap-and-dirty" system uses a single tracking device that measures the elevation angle of the model as seen from a single tracking location at a known ground distance from the launch pad. Two people are required—one to launch the model, and the other to track it in flight.

An elevation-only tracking device is shown in Figure 16-3. This is the simplest form of an elevation-only tracker. Estes Industries, Inc. sells an elevation-only tracking device, the AltiTrack® Altitude Finder, for those rock-eteers who don't want to make their own. An elevation-only tracking device is used with the "cheap-and-dirty" method and with the third system previously mentioned.

The setup for the cheap-and-dirty elevation-angle-only altitude determination method is shown in Figure 16-4. The launch pad is at L, the tracker at T, and the model rocket at its peak altitude at R. The distance LT is measured beforehand and should be about 300 feet (91.4 meters) or more—the more the better because the greater the distance LT, the better the overall accuracy of the system for high-altitude models.

When the model rocket lifts off from L, we assume that it flies vertically over the launch pad or that the distance from a point on the ground directly under the model to the tracker doesn't change much from the originally known LT distance. This is why the longer the distance LT, the more accurate this system.

The tracker operator follows the model rocket visually to the peak of its

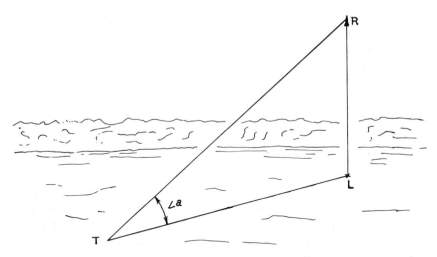

Figure 16-4 The geometry of the single-station elevation-only altitude tracking scheme.

flight, locks the tracking device with which he's been following the model, and reads the elevation angle that results.

The achieved altitude can now be calculated using simple trigonometry. We'll derive the equations here, but if you want to skip over the derivations and use just the information, that's up to you. You don't have to know *why* it works in order to make it work, although it certainly helps sometimes. It's not difficult to follow the derivations, and you might learn something in the process. Other people have.

By definition, angle RLT is a right angle. Therefore, according to the basic theorems of trigonometry, the *tangent* of the angle $\angle a$ is equal to the achieved altitude divided by the ground distance LT, as follows:

$$\tan \angle a = RL/LT \qquad (1)$$

Since we know the distance LT and the elevation angle a, we can rearrange the equation, collect all the known factors on one side, and come up with:

$$RL = LT \times \tan \angle a \qquad (2)$$

The tangent of the elevation angle can be found by looking in a tangent table such as is reproduced in Table 8. You can find similar tables in the back of any high school trig book.

Let's run through an example of elevation-only tracking. Suppose that the distance LT is 500 feet and, for this particular hypothetical flight, the

TABLE 8
Table of Tangents

Angle	Tangent	Angle	Tangent	Angle	Tangent
1	0.017	31	0.601	61	1.80
2	0.035	32	0.625	62	1.88
3	0.052	33	0.649	63	1.96
4	0.070	34	0.674	64	2.05
5	0.087	35	0.700	65	2.14
6	0.105	36	0.727	66	2.25
7	0.123	37	0.754	67	2.36
8	0.141	38	0.781	68	2.48
9	0.158	39	0.810	69	2.61
10	0.176	40	0.839	70	2.75
11	0.194	41	0.869	71	2.90
12	0.213	42	0.900	72	3.08
13	0.231	43	0.933	73	3.27
14	0.249	44	0.966	74	3.49
15	0.268	45	1.00	75	3.73
16	0.287	46	1.04	76	4.01
17	0.306	47	1.07	77	4.33
18	0.325	48	1.11	78	4.70
19	0.344	49	1.15	79	5.14
20	0.364	50	1.19	80	5.67
21	0.384	51	1.23	81	6.31
22	0.404	52	1.28	82	7.12
23	0.424	53	1.33	83	8.14
24	0.445	54	1.38	84	9.51
25	0.466	55	1.43	85	11.4
26	0.488	56	1.48	86	14.3
27	0.510	57	1.54	87	19.1
28	0.532	58	1.60	88	28.6
29	0.554	59	1.66	89	57.3
30	0.577	60	1.73	90	—

elevation angle of the model rocket when it reaches apogee is 32 degrees. The problem is solved as follows:

From trig table: $\tan 32° = 0.625$
$$RL = LT \times \tan \angle a$$
$$= 500 \times 0.625$$
$$= 312 \text{ feet}$$

The elevation-only method assumes that the model flies vertically over the launch pad so that the known ground distance LT doesn't change. Only

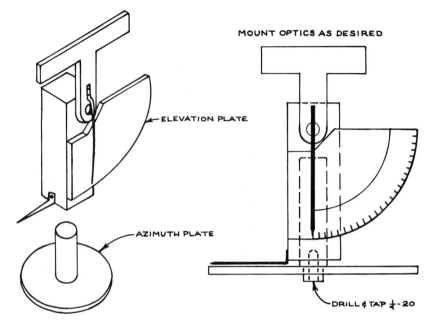

MOUNT OPTICS AS DESIRED

←ELEVATION PLATE

←AZIMUTH PLATE

DRILL & TAP ¼-20

Figure 16-5 Simplified sketch of an alt-azimuth tracking theodolite that can be made in most workshops.

a very few models will have this sort of perfect flight. Most will weathercock into the wind or otherwise deviate from the vertical flight path. If the model flies *toward* the tracking station, the distance LT becomes less and the elevation angle, assuming the altitude remains the same as for a vertical flight, becomes greater. Thus, the reduced altitude data will show that the model apparently flew to a higher altitude than was actually the case.

This shortcoming can be partly eliminated by using a very long baseline LT and by locating the tracker so it's crosswind from the launch pad. Then, when the model weathercocks, it doesn't fly toward or away from the tracker, and the distance LT doesn't change very much.

To fully account for the fact that a model rocket may not always fly vertically, a tracking system must be able to give an achieved altitude figure no matter *where the launch pad is located with respect to the trackers and no matter where in relation to the tracking stations the model rocket reaches apogee.* To meet these specifications, a very complex system of many trackers would be required. However, the specs can be met for more than 99 percent of all model rocket flights using two tracking devices capable of following the model in elevation *and* azimuth.

An azimuth angle is an angle measured in the horizontal plane.

Figure 16-6 The easy-to-build Triple-Track Tracker designed and perfected by Trip Barber.

A tracking device that will measure both elevation and azimuth angles is called a *theodolite*.

Many theodolites have been built and used in model rocketry. Some of them are very simple, while others have been complex. However, the complicated theodolites haven't produced any better accuracy or data than the simple ones.

The basic parts of a tracking theodolite are shown in Figure 16-5. This assembly may be threaded for mounting on any sturdy camera tripod. It may be made from wood, metal, or plastic, depending upon the material and shop facilities available. Inexpensive plastic protractors are perfectly adequate for use on model rocket tracking theodolites, provided reasonable care is taken to insure accurate assembly of the theodolite so that both elevation and azimuth axes pass accurately through the center of the protractors and that the pointers will retain angular accuracy as well. These basic theodolites will measure horizontal azimuth angle and vertical elevation angle to the NAR and FAI accuracy requirements of plus or minus a half degree.

Several kinds of optical tracking aids may be attached to this basic theodolite base. The simplest and most effective is nothing more than a straight piece of metal or wood 12 to 18 inches long. A small headless brad is driven into the stick at both ends. These provide two reference points with which to line up the tracker with the tracked object.

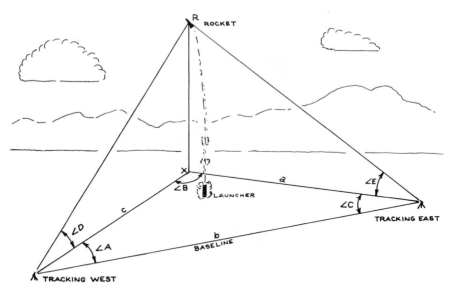

Figure 16-7 The geometry of the two-station tracking scheme.

This sort of open bead sight arrangement with no lenses or optical magnification has been found to be excellent and highly accurate as long as the model rocket, its smoke trail, or a puff of tracking powder at ejection can be seen with the unaided eye. Telescopes or rifle scopes provide a restricted field of view, and high magnifications actually work to the deteriment of model rocket tracking. If the model rocket can't be seen in the first place, telescopes aren't going to help a bit.

A somewhat more elegant sighting device is a mailing tube or body tube 2 inches in diameter and about 18 inches long with black thread cross hairs glued across both ends *and* two headless brads mounted atop the tube to provide an open field of view. The brads are used during the initial flight tracking, while the cross hairs inside the tube permit accurate aligning of the theodolite with the model as it slows down near the top of its flight.

The secret of tracking with this simple equipment is to watch the model with both eyes open, looking generally over the top of the tracker. Follow the model with your eyes as it ascends, keeping the tracker moving with the model. But don't try to zero-in on the model with the tracker at this stage because you don't need to. As the model slows down near apogee, *then* zero-in on it with the theodolite and get it right atop the brads or right in the cross hairs.

Models rockets 1/2 inch in diameter and 8 inches long have been tracked to altitudes of more than 1,800 feet with this sort of simple tracking equipment without any telescopes at all.

I don't recommend telescopes or rifle scopes be used on trackers. I've seen and used all kinds of trackers since 1958, and if I'm going to track I like to use simple open sights. I've got excellent distance vision—I pass an FAA pilot's physical exam every two years—and the chances are that you do, too, or that you wear glasses to correct your sight to normal distance vision.

If a model rocket can't be tracked, competition rules imply that it's not the fault of the trackers, but that of the builder of the model because the modeler didn't build a bird that was big enough or painted properly so that it could be tracked. In science and technology, it's no good if you build something whose performance exceeds the capabilities of the measurement system.

The theodolite system using azimuth and elevation angles requires two trackers positioned at the end of a measured baseline. The tracking situation is shown in Figure 16-7. For national and international competition the baseline must be at least 300 meters (984 feet 3 inches) long. The two tracking theodolites and the baseline may be set up in any relationship to the location of the launch pad. It is not necessary and, in fact, *highly undesirable* to have the launch pad located on the baseline. According to research carried out by J. Talley Guill in 1972 and confirmed by my own computer analysis of tracking situations and errors, the optimum location of the launch pad for best tracking accuracy is at an azimuth angle of 30 degrees from each tracker.

The two tracking locations or stations must have a clear view of the launch pad and of each other. The baseline between them should be horizontal, i.e., don't put one station on top of a hill and the other one down in a valley; put them on the same level with one another.

For the best view of the model in flight, position the stations south of the launch pad. Place one station southeast and the other southwest of the pad. These locations permit both stations to see the model in the full light of the sun. Don't position the stations so that one of them looks into the sun while tracking a model.

With the baseline measured and the stations located, the theodolites are set up at each station. Each theodolite should be leveled so that the azimuth protractor or measuring plane is horizontal.

Once both theodolites are *leveled*, the system is *zeroed* by sighting *both* theodolites directly *at each other* along the baseline. With the theodolites leveled and zeroed-in on each other, their azimuth *and* elevation dials or pointers are set to read *zero* azimuth angle and *zero* elevation angle.

Don't zero the theodolites on the launch pad. That doesn't work with this method.

Once this has been done, the tracking system is said to be *calibrated*.

We now have a tracking situation with a known distance between two stations plus, once the flight is made, an azimuth angle and an elevation angle from two tracking theodolites. This is more than enough data to compute the achieved altitude.

To understand how this is done, let's derive the equations used, referring to Figure 16-7 in the process.

Given: Distance *b*

Angle ∠A

Angle ∠D

Angle ∠C

Angle ∠E

Point X is an imaginary point on the ground directly beneath the apogee of the model in the air. The model rocket at apogee is at point R. We know the distance *b*; it's the measured baseline. So we need to compute the distance RX, which is the vertical altitude of the model rocket.

If we can compute the distance *a* or the distance *c*, we can theoretically locate point X on the ground. Then we can compute for the two vertical triangles to find the value of RX in two different calculations, checking one against the other to determine accuracy. These two vertical right triangles (R-X-West and R-X-East) are used to compute RX separately for each, and the value of RX from both is then averaged to produce a more accurate number.

The Law of Sines states:

$$c/\sin \angle C = b/\sin \angle B = a/\sin \angle A$$

Therefore:

$$\begin{aligned} c \ &= \ sin \angle C \ (b/\sin \angle B) \\ &= \ sin \angle C \ \{b/\sin[180 - (A + C)]\} \end{aligned}$$

Since R is directly above X by definition, the angle R-X-West is a right angle. We can therefore compute the western triangle as follows:

$$\tan \angle D = RX/c$$

Therefore:

$$RX = c \tan \angle D$$

Substituting for *c*, we get:

$$RX = \sin \angle C \tan \angle D\{b/\sin[180 - (A+C)]\}$$

The other vertical right triangle, the eastern triangle, is solved in an identical manner:

$$RX = \sin \angle A \tan \angle E\{b/\sin[180 - (A + C)]\}$$

The two values of RX thus obtained are then averaged. If either value deviates by more than 10 percent from this average altitude, it can be assumed that something went wrong. It means "track lost." But if both values of RX come within plus or minus 10 percent of the average, it's considered "track closed" and the achieved altitude data is considered to be official.

You don't have to memorize all this mathematical "banjo music" to make the system work for you. All you have to do is be able to read some tables, write some numbers into blanks, and do some common arithmetic. We've worked out lots of simple ways to do it. Let's start with the most difficult and proceed to the simplest.

Here's a tracking example:

Given: Baseline b = 300 meters
Tracking East azimuth ($\angle C$) = 23°
Tracking East elevation ($\angle E$) = 36°
Tracking West azimuth ($\angle A$) = 46°
Tracking West elevation ($\angle D$) = 53°

For one triangle we compute:

RX = $\sin \angle C \tan \angle D \{b/\sin[180 - (A + C)]\}$
  = $\sin 23° \tan 53° \{300/\sin[180 - (45 + 23)]\}$
  = $0.391 \times 1.33 \times 300/\sin 68°$
  = $0.391 \times 1.33 \times 324$
  = 168 meters (551 feet)

Solving the other triangle by the same method yields RX = 167 meters (548 feet).

The average of these two altitudes is 168 + 167 divided by 2 which equals 167.5 meters which is the average altitude.

In applying the "10 percent Rule," the average altitude is rounded-off using the rule-of-thumb, "Keep it even"—i.e., if the average altitude ends in a decimal tenth of 4 or less, the decimal is dropped; if it ends in a decimal number of 6 or more, it's rounded-up to the next whole number; if the decimal tenth is five, it's rounded down if the number immediately to the left of the decimal is even; if the decimal tenth is 5 and the number immediately to the left of the decimal point is odd, the number is rounded-up to the next whole number.

In our example, the rounded number of the average altitude would be 168 meters.

Both computer altitudes—168 and 167—fall within 10 percent of the average altitude of 16.8 meters. Therefore, the track is considered good.

This system was originally worked out back in 1958 by Arthur H. Ballah and Grant R. Gray. It was refined into a very rapid system in 1960 by John S. Roe who devised a table giving sine and tangent values of angles *plus* a column for the value of 300/sin∠B. See Table 9. This method is very fast and requires only a four-function pocket calculator and a pencil and paper.

Naturally, the next thing to do was to put the whole tracking situation into a large high-speed digital computer and instruct it to solve for every possible combination of azimuth and elevation angles. This programming was first done by J. Talley Guill at Rice University in 1969 and produced the first of a series of precomputed altitude tables which were used at national meets for many years.

The advent of the microprocessor, minicomputer, and pocket programmable calculator resulted in even simpler data reduction without the need to do anything more than enter the angular data into the computer and instruct it to solve both triangles, average the results, round-off the average altitude, apply the 10 percent rule, print out the results, and report whether or not the track closed. Gary Crowell was the first one to write a program that was successfully used on the field at NARAM-20 in Anaheim, California, in 1978. Gary Crowell and I subsequently collaborated to write a simple data reduction program in BASIC, which is shown in Appendix VII.

A third tracking system was developed by J. Talley Guill who believed it should be possible to have a tracking system using several simple elevation-only trackers like the one shown in Figure 16-3 or the Estes AltiTrak®. He perfected his system in the summer of 1972 and it was checked out for suitability and accuracy by comparing it side-by-side with the standard two-station azimuth-elevation system just described in detail above. Three elevation-only trackers are required, and they must be located on a single baseline as shown in Figure 16-9 with one tracker at each end and the third tracker precisely in the middle of the baseline. The data reduction operations and tables for the three-station elevation-only Guill system are detailed in Appendix VI. It works and is a simpler but perfectly accurate system for those clubs that don't want to build theodolites and prefer to buy 3 AltiTraks® instead.

When you set up tracking stations, you've got one of the elements of a model rocket range. We'll discuss all the aspects of a full model rocket range in the next chapter. But even using a single elevation-only tracker, you must have some means of communication between the tracker and the launch pad if for no other reason than to inform the tracker of an imminent launch so that the model's lift-off doesn't come as a complete surprise.

The simplest range communications system I ever saw was used by the Czechoslovaks at the First International Model Rocket Competition in Dubnica-nad-Vahom, Czechoslovakia, in May 1966. One hand-held flag is at the launch pad, and the other at the tracking station. If more than one tracking station is used, there's a flag at each one. When a model is ready to launch,

TABLE 9
Altitude Calculation Table or Two-station Theodolite System

Use this table for computing the altitude of models using elevation and azimuth angles from two tracking stations on a base line of 300 meters, or 984.24 feet. $V = 300/\sin B$.

Angle	Sin	Tan	v
1	0.0174	0.0174	17189.6
2	0.0349	0.0349	8596.11
3	0.0523	0.0524	5732.2
4	0.0698	0.0693	4300.7
5	0.0872	0.0875	3442.1
6	0.1045	0.1051	2870.0
7	0.1219	0.1228	2461.6
8	0.1392	0.1405	2155.6
9	0.1564	0.1584	1917.7
10	0.1736	0.1763	1727.6
11	0.1908	0.1944	1572.3
12	0.2079	0.2126	1442.9
13	0.2249	0.2309	1333.6
14	0.2419	0.2493	1240.1
15	0.2588	0.2679	1159.1
16	0.2756	0.2867	1088.4
17	0.2924	0.3057	1026.1
18	0.3090	0.3249	970.8
19	0.3256	0.3443	921.47
20	0.3420	0.3640	877.14
21	0.3584	0.3839	837.13
22	0.3746	0.4040	800.84
23	0.3907	0.4245	767.79
24	0.4067	0.4452	737.58
25	0.4226	0.4663	709.86
26	0.4384	0.4877	684.35
27	0.4540	0.5095	660.81
28	0.4695	0.5317	639.02
29	0.4848	0.5543	618.80
30	0.5000	0.5774	600.00
31	0.5150	0.6009	582.48
32	0.5299	0.6249	566.12
33	0.5446	0.6494	550.82
34	0.5592	0.6745	536.49
35	0.5736	0.7002	523.03

Angle	Sin	Tan	V
36	0.5878	0.7265	510.39
37	0.6018	0.7535	498.49
38	0.6157	0.7813	487.28
39	0.6293	0.8098	476.70
40	0.6428	0.8391	466.72
41	0.6561	0.8693	457.28
42	0.6691	0.9004	448.34
43	0.6870	0.9325	439.88
44	0.6947	0.9657	431.87
45	0.7071	1.0000	424.26
46	0.7193	1.0355	417.05
47	0.7313	1.0723	410.20
48	0.7431	1.1106	403.69
49	0.7547	1.1504	397.50
50	0.7660	1.1918	391.62
51	0.7715	1.2349	386.03
52	0.7880	1.2799	380.71
53	0.7986	1.3270	375.64
54	0.8090	1.3764	370.82
55	0.8192	1.4281	366.23
56	0.8290	1.4826	361.87
57	0.8387	1.5399	357.71
58	0.8480	1.6003	353.75
59	0.8572	1.6643	349.99
60	0.8660	1.7321	346.41
61	0.8746	1.8040	343.01
62	0.8829	1.8807	339.77
63	0.8910	1.9626	336.70
64	0.8988	2.0503	333.78
65	0.9063	2.1445	331.01
66	0.9135	2.2460	328.39
67	0.9205	2.3558	325.91
68	0.9272	2.4751	323.56
69	0.9336	2.6051	321.34
70	0.9397	2.7475	319.25
71	0.9455	2.9042	317.29
72	0.9511	3.0777	315.44
73	0.9563	3.2709	313.71
74	0.9613	3.4874	312.09
75	0.9659	3.7320	310.58
76	0.9703	4.0108	309.18
77	0.9744	4.3315	307.89

Angle	Sin	Tan	V
78	0.9781	4.7046	306.70
79	0.9816	5.1445	305.62
80	0.9848	5.6713	304.63
81	0.9877	6.3138	303.74
82	0.9903	7.1154	302.95
83	0.9925	8.1443	302.25
84	0.9945	9.5144	301.65
85	0.9962	11.4300	301.15
86	0.9976	14.301	300.73
87	0.9986	19.081	300.41
88	0.9994	28.656	300.18
89	0.9998	57.29	300.04
90	1.0000	∞	300.00

the flagman at the launch pad raises his flag. The trackers see this; when they're ready to track the model, they raise their flags. When the flag at the launch control point drops, the model is launched. The trackers then write down their angles and send them back to the launch control point by a runner—the Czechs used motorcycle couriers.

Although Citizen's Band walkie-talkie radiotelephones have been used to communicate between the launch area and the trackers, this communications method really hasn't worked out because of interference in addition to the absence of two-way simultaneous or "duplex" telephone-like communication wherein a tracker can call a hold if necessary over the countdown from launch control.

The best range communication system is still the one originally developed on the very first model rocket range, Green Mountain Proving Ground, Denver, Colorado, in 1958: the land-line telephone network.

Land-line telephone systems *are* expensive because they require lots of wire; a way to string out, wind in, and store that wire; and telephones. Some clubs have gotten the support of their local telephone company to obtain telephone handsets and wire. Building a land-line telephone system otherwise means locating wire and surplus telephone units through electronics supply houses listed in the Yellow Pages of the telephone directory, putting it all together, and building the necessary wire-winding reel to store the wire on— unless you have a permanent launch site and can bury the wire in the ground on a permanent basis (few clubs enjoy the exclusive use of a large piece of land for their flying activities).

You will have to find about 2,000 feet of plastic covered "twisted pair" or two-conductor telephone wire. Using a single wire with the ground as the

NAR ValSun SECTION   FLIGHT **D**ATA SHEET   Date: _____

NAME: _____ NAR #: _____

MODEL NAME: _____ MOTOR TYPE(S): _____

LAUNCH ALLEY: _____SAFETY CHECK BY: _____

TRACKING DATA: Baseline length: _____ Recorded by:_____

Tracking East: Azimuth_____° Elevation: _____°

Tracking West: Azimuth: _____° Elevation: _____°   LOST:☐

DATA REDUCTION: DR by: _____

Tracking East: Azimuth____° sin_____ ①Elevation: ____° tan ____ ②

Tracking West: Azimuth____° sin_____ ③Elevation: ____° tan____ ④

ADD azimuths for B: _____° Table Value V for B: _____ ⑤

MULTIPLY: _____ ⑤ x _____ ② x _____ ③ = _____

MULTIPLY: _____ ⑤ x _____ ④ x _____ ① = _____

                                   ADD: _____

THIS IS AVERAGE ALTITUDE: DIVIDE BY 2:   _____

10% RULE CALCULATION:

Average altitude: _____   If RESULT is LESS than
                                          HIGHEST of the two
ADD 10% of average altitude: _____   computed altitudes, OK

RESULT: _____
                                          _____OK   _____LOST
      -

Average altitude: _____   If RESULT is MORE than
                                          LOWEST of the two computed
SUBRACT 10% of Average Altitude: _____ altitudes, OK.

RESULT: _____   _____ OK   _____LOST

NOTES:

Figure 16-8 Typical flight data sheet used to record and reduce tracking data for a model flight.

return circuit doesn't work any better today than it did over a century ago with the first, primitive telephones.

The simplest telephones available are "sound-powered" handsets such as those made by Wheeler and other firms for industrial use. Three handsets are required for a two-station tracking system range—one at each station and one at launch control. No batteries are required. Simply clip the separate wires from the headset to the separate wires of the twisted-pair field wire. Sound-powered telephones aren't very loud, but they're adequate.

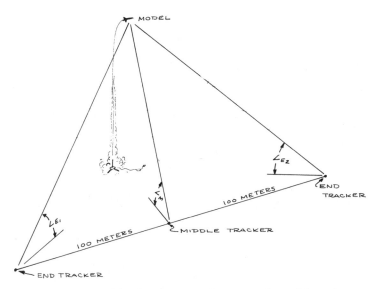

Figure 16-9 The geometry of the three-station elevation-only tracking scheme. See Appendix VII.

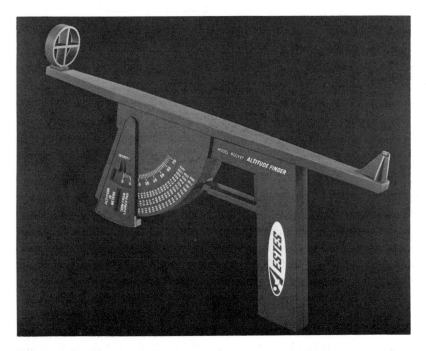

Figure 16-10 The Estes AltiTrak® elevation-only tracking device. (Estes Industries, Inc.)

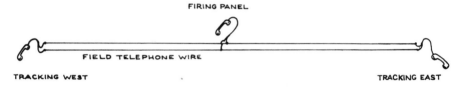

Figure 16-11 Wiring diagram for a simple sound-powered range telephone system.

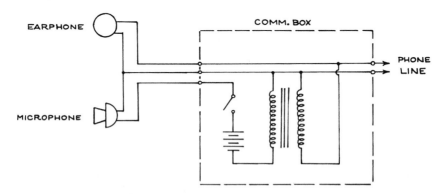

Figure 16-12 Wiring diagram of the communications box described in the text for the standard land-line range telephone system. See text for parts description.

To build a regular land-line telephone system, you're going to have to be *very* resourceful to locate telephone handsets. You don't need the entire telephone, only the hand-held receiver and transmitter known as the "handset." It should have a magnetic receiver and a carbon microphone element. If you can't talk the local telephone company out of three or four old handsets, you'll have to check with local electronic supply stores, surplus stores, and junk stores. Surplus army field telephones work great and require no additional work on your part except to hook them to the paired field wires, but you'll be very fortunate to be able to find three of these that are inexpensive enough.

Once having obtained your handsets (or even telephone operators' "head-sets" with carbon mike elements) you'll have to build a communication box for each telephone handset. All the parts can be obtained at any well-stocked

Radio Shack store, and Radio Shack part numbers are given here. A comm box wiring diagram is shown in Figure 16-10. The box contains the following:

1. a 1.5-volt Size D flashlight battery (23-466) to provide excitation to the carbon microphone element.
2. a battery holder (270-403) for the Size D flashlight battery.
3. a spring-loaded normally open "push-to-talk" switch (275-609) so that the handset mike doesn't pick up background sounds or transmit unless the station operator pushes the switch to do so.
4. an "isolation transformer" (273-1386 or any matching or isolation transformer with turns or impedance ratios up to 5:1) wired as shown to keep the battery voltage out of the handset receiver and off the land-line wires.
5. a three-wire input jack (274-141) for the telephone handset unless you wire the handset permanently into the box.
6. a handset input plug (274-139) for the wire coming from the handset, unless you wire the handset permanently into the box.
7. a jack for the output of the box (274-252 unless you use two binding posts (274-662) mounted on the box to which you'll attach each of the wires in the field wire.
8. a field wire plug (274-139) on the end of the field wire to plug into the jack on the box, unless you use two binding posts as above to attach the field wire directly to the box circuitry.
9. an aluminum chassis box (270-236) in which to house all the parts.
10. miscellaneous AWG-22 stranded wire, rubber grommets, plastic wire clamps, nuts and bolts, and other pieces.

At each tracking station and at launch control is a handset and a comm box. The field wire is simply attached to each comm box. The comm box at launch control will have the ends of both lines out to the trackers attached to it.

A land-line telephone system such as this is rugged, long-lived, easy to maintain, easy to set up, easy to operate, easy to take down, and capable of being wired up to public address systems and other sound equipment. Such systems have been in weekly use on model rocket ranges for as long as 15 years, and they show no signs of giving up yet.

Thus, while a land-line telephone system is expensive, hard to find the parts to build, difficult to build, and not compact, it's a one-time expense for a club and will last for a very long time indeed.

With a land-line telephone system and trackers, you're now well on your way to having a complete model rocket range on which you could even hold a regional competition!

17

Model Rocket Ranges

Although you may begin flying model rockets in an open field with one or two other people, you'll soon be joined by others who want to fly there, too. You can have a lot more fun and learn more when you fly with other model rocketeers. However, if many models are flown and if they are to be tracked for altitude, some additional equipment will be required. And you'll have to have some form of organization to maintain safe operations and to prevent confusion.

When most people visit a model rocket range for the first time, they're impressed by its organization and safety control. These model rocket range characteristics did not evolve haphazardly; model rocket range operations were carefully designed because *safety is no accident.* The first time model rocketeers got together to fly, it was at White Sands Missile Range in New Mexico, and we were all aware from our professional rocket work that range organization was absolutely necessary. We'd been thoroughly indoctrinated with the safety procedures used in flying real rockets and guided missiles. Few model rocketeers realize the tremendous debt they owe to two men who pioneered the flight safety practices of full-scale rocketry and whose policies were adopted by model rocketeers. They are Herbert L. Karsch and Nathan Wagner, the flight safety officers and engineers at White Sands back in the 1950s. Much of what you're going to read about in this chapter was adopted straight from White Sands where all of us got our early training and experience in rocketry because at that time it was the only place to do it. It's largely because of Karsch and Wagner that model rocketry, since its inception in 1957, has conclusively proven itself to be safer than model airplane flying and safer even than bicycle riding.

The Number One Safety Rule adopted from professional rocketry is: The Range Safety Officer's word is as the word of God, and *nobody* can override a safety decision made by a Range Safety Officer (RSO). If the RSO says, "No," that's the end of it. Either do what he says or go someplace else.

At White Sands, and later at Cape Canaveral, Vandenberg Air Force Base, Point Mugu, Eglin Air Force Base, Fort Churchill, Tonopah, Green River, and NASA Wallops Flight Center, the Range Safety Officer is supreme. He or his trained crew checks every rocket vehicle before flight. He alone determines whether or not the rocket is safe to launch. He alone determines that the launch range is ready and in a safe condition. He gives the final safety clearance before launch. He even destroys the vehicle in the air if it becomes unsafe.

The basic rocket range philosophy for the big ones and the little ones is: Don't take chances. It isn't worth it. That's one reason why rocketry is still "the world's safest business."

Those of us who started model rocketry profited from the experiences of the professional range safety people and adopted as many of their practices as practical. The only thing we don't do in model rocketry is destroy the model in midair if it becomes unsafe, although I've sometimes wished we had such a procedure. I once remarked on this to the recently retired Director of NASA Wallops Flight Center, Robert Krieger, who replied as we watched a model rocket lift off, "Well, if you'll get me a 12-gauge shotgun . . ."

While adopting professional range safety procedures for model rocket ranges, we also borrowed many other aspects of full-scale rocket flight testing. Range operations procedures were one such aspect.

Although model rocketry is a very individualistic hobby where a person can use creativity to the utmost within the bounds imposed by the real universe in the design and construction of model rockets, it's also a team effort on the model rocket range where everyone works together.

A model rocket range is a little Cape Canaveral. It's run the way the Cape is. If anyone who had a rocket to fly were allowed to set up his launch pad anywhere at the Cape and launch whenever he wanted to, there would be chaos. The same holds true for a model rocket range. It's not only fun to work together in a professional manner, but operating in this way helps tie our hobby closer to the real thing.

The first requirement for a model rocket range is a large open space of land. For models flying up to 1,000 feet (300 meters), an ordinary school football field is adequate if it isn't hemmed in by too many rocket-eating trees. If you're going to fly large models or do a lot of glide recovery work, you'll have to get a larger field. Model rocketeers face the same problems as model aviators: Where can we find a flying field that's large enough? In the western and rural parts of the United States, model rocketeers usually don't have too much of a problem finding a flying site if they're willing to drive 10 or 20 miles. Somebody always has a friend or relative with a farm or ranch land somewhere. But this isn't true if you're among the 90 percent of the American people who live in congested urban and suburban areas. As the old maps used to point out, here there be complaining neighbors, uncooperative fire marshals, and/or insolent bureaucrats or politicians. But these people can usually be transformed into friends by careful diplomacy.

There are no pat answers to the problem of finding a place to fly. But you should always have the *written* permission of the owner of the land. In

Figure 17-1 A model rocket range is a place of organized activity where modelers can fly their rockets in safety and gain more information about model performance.

my own experience, it's always been easier and simpler to get permission to use privately owned land such as a farm meadow. Getting permission to use public land often means dealing with the bureaucracy, which can become complicated and frustrating. However, many model rocket clubs have been sponsored by local civic and youth groups, and this paved the way for getting permission to use parks and recreational areas.

The most important single thing that will help a group of young model rocketeers obtain permission to use a flying site is an adult "advisor" who may be a member of the club and who's willing to act for you. An adult will know how to approach local authorities and can speak as a tax-paying, voting citizen. Believe me, this helps! An adult spokesman can also serve as the club's Range Safety Officer, club advisor, or club sponsor. More about that in the next chapter.

Any flying field that you select should be reasonably free of trees, of course, because regardless of what the botanists say, all trees are rocket-eating trees. It should also be clear of high-voltage electric transmission lines; any model that gets hung up in a power line should be left there because you could get killed trying to remove it.

A flying field should also be away from heavily traveled roads and high-speed highways, parkways, turnpikes, freeways, and interstates. A model rocket won't damage an automobile, but if a parachuting model or glider wafts down in front of a driver speeding along in heavy traffic, it's going to startle him and may cause an accident.

Stay away from airports, airport runway approach zones, and other areas around which there may be low-flying aircraft. In addition to being a model rocketeer, I'm also a pilot, and I can assure you that a pilot is very busy and has a lot to do during takeoff and landing at an airport without also having to worry about model rockets. It's practically impossible to hit an airplane in flight with a model rocket—one chance in 50 billion flights with a 95 percent confidence factor. The Department of Defense has spent billions of dollars to develop guided missiles that will deliberately hit airplanes, and a little unguided model rocket isn't going to achieve what an expensive, complex, and sophisticated Chaparral, Hawk, or Redeye SAM is designed to accomplish.

If your small local airport is used mostly by private aircraft and is the only place around with enough room for flying model rockets, fly there only with the permission of the airport director, and get permission every time you want to use the field. Flying model rockets directly on the airport itself is probably safer that flying them a mile away from the airport. National model rocket competitions have been conducted with perfect safety from the middle of active first-class airports (Lawrence G. Hanscom Field in Bedford, Massachusetts, for example), so there's really not much to worry about if your RSO stays alert.

The layout of an *idealized* model rocket range is shown in Figure 17-2. Note that I emphasized the word *idealized*. Someday I'd like to fly on such an ideal model rocket range. I've been on some very good model rocket ranges, but every location is somewhat less than ideal.

Note that the launch area itself is located at or near the center of the field. This makes flying and recovery independent of the wind direction. However, if the prevailing wind is predominantly from one direction, you can offset the location of the launch area to the *upwind* side of the field to provide more downwind recovery area.

The tracking stations are located on the south side of the field in positions where the trackers don't have to look into the sun when following models in flight. Note that the trackers have a clear and unobstructed view of the launch area at all times.

The launch area normally faces north so the Range Safety Officer, the Range Control Officer, and other model rocketeers have the sun *at their backs* when flying. You may not think this is a very important point until you've been on the field for the whole day and are suffering sunburned eyeballs and tonsils because somebody set things up so that you always looked into the sun. A couple of national meets were inadvertently set up with the launch

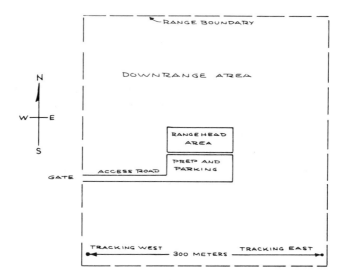

Figure 17-2 The layout of an idealized model rocket range.

area facing into the sun, and this resulted in five miserable days of model rocket flying and some very painful sunburns.

There should be only one entrance to the flying field. A sign should be posted at the entrance to let people know what's going on and to advise them to stay alert if they come on the field to watch. The wording of the following sign is recommended:

PODUNK MODEL ROCKET CLUB
FLYING FIELD
Model Rockets in Flight
PROCEED WITH CAUTION

The sign shouldn't say such things as, "Danger! Look Out For Falling Rockets!" or other such things suggesting that you're not conducting a very safe activity.

Cars should be parked so they don't block the line of sight from the trackers to the launch area. If possible, cars should be allowed on the field because it's nice to be able to work out of a trunk or the back end of a station wagon. I've been on fields where cars weren't allowed and we had to carry everything for more than a hundred yards into the launch area; it was miserable.

The launch area is usually called the rangehead area. Figure 17-3 shows the layout of a typical modern rangehead area. It's called the Alley Rangehead Layout, and it's been in use since 1965. It replaced an older system that used

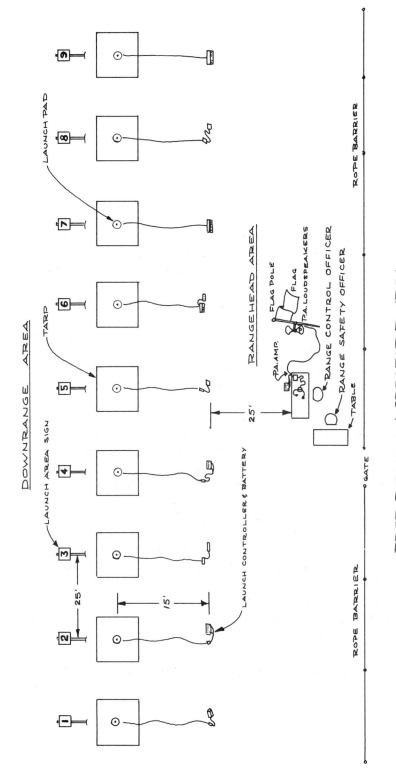

Figure 17-3 The layout of the rangehead area of the Alley range system.

racks of launchers on saw horses, although you'll still see "torture racks" used by some of the older, less progressive clubs.

The rangehead area is separated from the spectator and prep areas by a simple rope barricade. There's an open gate in the center. The rope barricade keeps spectators out of the rangehead area. Everyone who enters and leaves the rangehead area should use the gate and not jump the barricade. Jumping the barricade implies lax range discipline. If it's a big rangehead area, put in extra gates to keep modelers from jumping or ducking under the barricade.

The barricade isn't very sturdy. It's very simple, and basically symbolic in nature. Often a gasoline station will give you some of its used or extra bannered ropes, the kind with colored triangular plastic pennants hanging from them. These are colorful and attractive, and they liven up the appearance of the range. They are made into a barricade by simple wooden dowels or steel fence stakes driven into the ground to support them. There should be a post every 20 feet or so.

Fifteen yards in front of the barricade are the launch areas. Each is marked by a large sign giving the launch area number. The easiest way to make launch area markers is to buy 12 or 18 white plastic wastebaskets; turn them upside-down and paint numbers on two sides of them. Or you can make signs that stand up where they're easier to see.

There should be 15 to 20 feet (5 paces) laterally between launch areas.

If you're flying on a grassy field or one with dry grass in the rangehead area, each launch area should have a canvas tarp at least 5 feet square on which a launch pad can be set up. The tarp serves two purposes. It keeps you from getting your pants or knees dirty when you're on your knees working around your launch pad hooking up your model. It also prevents grass fires. A grass fire is no fun. On March 31, 1967, I was on hand when our club accidentally burned off an 11-acre grass field. All safety precautions were in effect, but a glowing piece of igniter landed in the grass next to a launch pad. A 15-mile-per-hour wind was blowing, and by the time the RSO saw the fire a few seconds after launch, it was out of control. The club purchased a tarp for each launch area, and we never had another grass fire in the remaining six years I was with that club.

An additional precaution to take when the grass gets dry is to bring a 5-gallon garden sprayer filled with water to the range and put it in the rangehead area. If things are really dry that day, appoint a Fire Guard whose job it is to stand with the water sprayer just behind the modeler who's launching. The Fire Guard *never* takes his eyes off the launch pad, even when the model's launched. It's his job to *watch that launcher* and to get out to the launcher with the water if by chance a conflagration starts. (Be careful who you assign as Fire Guard. Some people are exuberantly enthusiastic about the job, especially when it comes to completely soaking down a competitor's launch pad on the slightest excuse.)

Some people believe it's necessary for safety purposes to set up rope barricades to separate the launch alleys and launch areas. This is wrong. *Barriers and barricades to separate launch alleys and areas can be extremely dangerous because they restrict personal movement in the rangehead area and give people a false sense of protection.* Rangehead barriers and barricades channel people into specific areas, which can be dangerous if something does go wrong during a flight. On any model rocket range, you want to be able to *move* and *move fast* if something does go wrong; you don't want to find a barrier in your way. I've pioneered the open rangehead sort of Alley Rangehead Layout, and I've flown on Alley ranges with barriers everywhere. Once you discipline yourself to the open Alley, you feel hemmed in on a barriered Alley range. If the RSO and RCO don't have the sense to call people for infractions of rangehead discipline or the guts to take disciplinary action when necessary, all the rope barriers in the world aren't going to make any difference.

Two portable folding tables are set up next to the barricade gate as shown in Figure 17-3. At one of these tables the Range Safety Officer (RSO) presides. Every model that's to be launched is taken into the rangehead area and presented to the RSO at his table. During contests, this RSO table is also used for working up flight cards. There may be a weighing scale on the RSO table if there's a contest going on.

The other table provides the Range Control Officer (RCO) with a place to work. He sits there in command of the entire range. The RCO has a *loud* battery-powered public address system and microphone. He also has communications with the tracking stations, if any, by means of the land-line telephone network.

There are several battery-powered public address or PA systems on the market today. Years ago, we had to build them ourselves. You should have at least a 10-watt PA system capable of working from a 12-volt 4.5 ampere-hour battery or, if you want to carry it, an auto battery. Make a few modifications to the PA system so the land-line telephone network can plug into it through the "aux," "phono," or other "high-level" input jack. If you put the land-line telephone net on the PA system, this will mean that the RCO can use the telephone handset instead of the mike for announcements, etc. It will also mean that *everybody* on the range will be able to hear the conversation with the trackers. The trackers will also be able to hear each countdown because the RCO uses the telephone handset which is also plugged into the PA system. It also allows people to hear the tracking stations reporting in to state their readiness or read their angular data. This is *very* important in contests. It also permits the range to be run by the RCO without requiring additional manpower for communicating with the trackers. Some clubs will keep their PA system separate from the telephone network; many of these organizations still use the torture rack system and are run like miniature

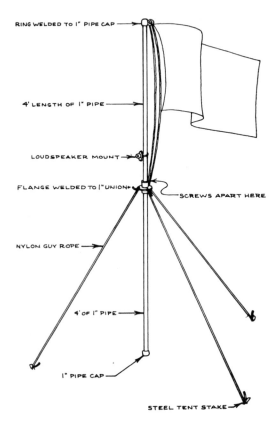

RING WELDED TO 1" PIPE CAP

4' LENGTH OF 1" PIPE

LOUDSPEAKER MOUNT

FLANGE WELDED TO 1"UNION

SCREWS APART HERE

NYLON GUY ROPE

4' OF 1" PIPE

1" PIPE CAP

STEEL TENT STAKE

Figure 17-4 Sketch of a simple portable flag pole and loudspeaker mount.

dictatorships where one or two people enjoy being absolute rulers of the range, complete with secret communications systems.

There should be *two* PA system loudspeakers mounted on a range flagpole near the RCO table as shown. Use the 10-watt metal horn speakers. Point one toward the rangehead area so that modelers can hear their countdowns on the PA system. Point the other one back toward the spectator and prep area so that everyone else can hear what's going on. This positioning is important because it lets everybody get the word about what's happening.

A good range should also have a flagpole, and it should be portable. A typical portable flagpole is shown in Figure 17-4. It can be made from pipe and other common hardware store items. It can also be made using two aluminum TV masts. Guy ropes are necessary, and they should be positioned such that they don't get in the way and that people can't trip over them or accidentally run into them. The PA loudspeakers can be mounted on the

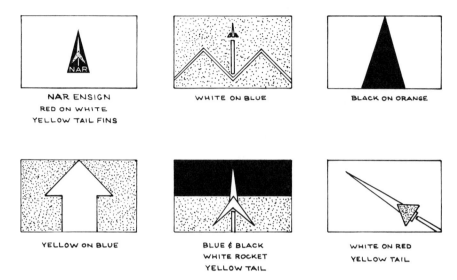

NAR ENSIGN
RED ON WHITE
YELLOW TAIL FINS

WHITE ON BLUE

BLACK ON ORANGE

YELLOW ON BLUE

BLUE & BLACK
WHITE ROCKET
YELLOW TAIL

WHITE ON RED
YELLOW TAIL

Figure 17-5 Some typical range flags of various NAR Sections.

flagpole either at the mid position or, preferably, at the top. Pulleys and blocks aren't necessary for the flag hoist; ordinary eyes will work fine.

You should fly our national colors with your club's own flag underneath. Or just fly your club flag. A flag is a very important part of a model rocket range. It draws attention to the fact that there's something going on in the field. It locates the center of the action. It's also a wind indicator. After a long trek back from recovering your model in the boondocks, it shows you where the rangehead area is so you don't get lost. The flag idea is something else borrowed from White Sands, which adopted it from rifle and artillery ranges where a red flag is always flown when the range is in operation.

At regional meets, provide flagpoles at 10-foot intervals along the rope barricade for other clubs to fly their flags. An excellent contest flagpole can be made from a 16-foot length of 2 × 2 wood with a screweye at the top, a cleat at the middle, and a flag hoist made with light curtain cord. These flagpoles must be emplaced the day before the meet starts by digging a post hole 4 feet deep for each flag pole, and emplacing the flag poles for the entire contest. They're easily pulled out of the ground right after the contest is over.

The equipment discussed thus far will cost about $200, depending upon how much can be located in the junk piles of members' workshops, how much can be built by handy members, and how much of it can be obtained or scrounged from local donors or sources.

Be sure to design the equipment so it breaks down into small elements that are easy to carry and pack away for transportation and storage in footlockers, specially made wooden boxes, or other containers. Make your range equipment as portable and compact as possible so that no more than two people are required to lift or carry any box or container. Design it so it will fit into the back end of a station wagon.

By all means, build your range equipment so that it is *rugged*. It will take a beating. If you build it with this in mind, you won't have to spend a lot of time repairing and maintaining it. Good range equipment will last for a decade or more of hard use with only occasional maintenance necessary.

How does a range such as this operate?

The first order of business when everyone arrives at the announced start time—30 minutes prior to the start of flying time—is to set up the range. Actually, as few as six people can set up a complete model rocket range such as we've been discussing here. However, the more people there are, the faster it goes together. Nobody should be permitted to start prepping models or setting up personal equipment in the rangehead area until the entire range is set up and checked out. If necessary, insure participation by pulling assignments out of a hat if a call for volunteers doesn't result in enough help.

The range can be operated without trackers by one person, a combined RSO–RCO, if manpower is short. The complete range with trackers requires four people—RSO, RCO, Tracking East, and Tracking West.

The Range Safety Officer performs safety-checking and control duties as described earlier. The RSO should be at least 18 years of age. Why? Occasionally, the RSO is going to have to make a difficult safety decision that somebody else won't like, so the RSO should be a person who can stick to a decision, often under considerable pressure from adults and parents. Therefore, long and hard experience has proven that only a person over 18 should assume the position, duties, and responsibilities of an RSO. Notice that the gender of the RSO hasn't been mentioned; a mother, older sister, or experienced woman model rocketeer has often proven to be a strict and competent RSO fully capable of making even the technical safety decisions required.

The Range Control Officer can be any club member. Although the job of the RSO is range safety, the RCO is in charge of operating the range itself. The RCO selects the next model to be launched, talks to the tracking stations, gives countdowns, records data, and keeps a running commentary going on the PA system. Although in the crunch of a heavy contest, a steady hand and a cool head are required of the RCO, it's a good idea for all club members to take a turn at this position, even those who are shy and quiet. Many people have learned how to handle themselves in public and how to keep calm in a hassle because of their training and experience as RCOs.

Two trackers are required if tracking stations have been set up—or three trackers if the three-station elevation-only system is used. There is no need

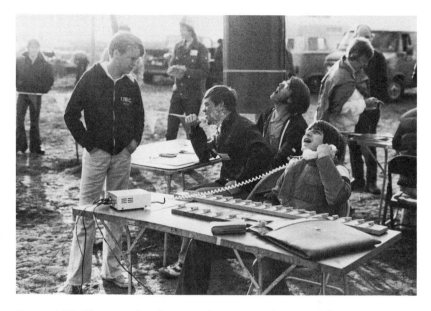

Figure 17-6 The rangehead area is always one of purposeful activity with the RSO and RCO on duty while modelers process their models and timers clock flight durations during a contest.

to set up or man the tracking stations if no tracking is to be done; the range then operates as a sport or duration range. However, to give people experience and training in tracking, the trackers should be set up and manned for at least two 30-minute shifts during the flying session, even though there isn't a contest or other requirement for tracking that day.

The most important thing is to get your club members qualified, experienced, and up-to-date on *all* range positions. Adults should be qualified as RSOs, RCOs, and trackers. Young members should be trained and qualified as RCOs and trackers. New members should be encouraged to try a new range position each flight session, and older members should devote two or three 30-minute shifts of the flying session to the three range positions.

Once the range is set up and ready to go, everybody should sign up on a "range manpower sheet" such as the one shown in Figure 17-7. Everyone should take at least one range position during one shift during the day. *The range should not be opened for flying until all positions on all shifts are filled on the Range Manpower Sheets.* If you don't do this, you'll end up with a couple of people doing nothing but running the range all day, and this isn't fair to them or to the other members who *should* be learning all the jobs.

Everyone who's going to fly picks a launch area and sets up his own

RANGE MANPOWER

Date: _____

Contest: _____

Time	Range Safety	Range Control	Track E	Track W	Return	Timers	Fire Guard	D	R
1:30-2:00									
2:00-2:30									
2:30-3:00									
3:00-3:30									
3:30-4:00									
4:00-4:30									
4:30-5:00									

Figure 17-7 The Range Manpower Sheet developed for use on the Alley type model rocket range. Everybody signs up for a half-hour shift during the afternoon.

launch pad, electrical ignition system, and battery as shown in Figure 17-6. The procedure where everyone has his own launch pad and ignition-battery system has proven itself to be superior to other forms of range operation because it stresses *individual responsibility*. Each person is responsible for his own equipment. Since each member builds, operates, uses, repairs, and maintains his own GSE, it's his own fault if it doesn't work. And it's usually working properly when the next flight session rolls around. It also means that one or two people aren't stuck with the nasty job of keeping a lot of launch equipment in operating condition, a maintenance task that can be a great deal of work.

Often two or more model rocketeers—especially families—share the same launch area because there's plenty of room at each area to set up two or three launch pads. There are seldom any conflicts in doing this because rarely do two or more modelers want to launch at the same time out of the same launch area. When it does occur and if it can't be resolved by the modelers themselves, the RCO can step in and assign priorities. (We discovered, however, that four Stines required two launch areas.)

There are two cardinal rules in the rangehead area, and these are important for members to remember and for the RCO to enforce:

1. Never cut across the launch alleys. You could trip over somebody else's ignition wires, and this makes them very unhappy.
2. Always approach the launch areas at right angles to the rangehead area. Keep your head up and listen for any countdown on the PA system. The RCO will be watching to ensure that you don't walk into a launch area that's in its terminal countdown, but you shouldn't depend on the RCO for your personal safety. Keep your eyes open and your head screwed on tightly.

Occasionally, when a large model powered by Type D, E, or F motors is launched, the RCO will ask that the adjacent launch areas be vacated during the terminal count.

When you've prepped your model in the prep area and are ready to fly, bring your model to the rangehead area through the gate and present it to the RSO for safety check. This is *always* done. If your model is to be tracked and you have a flight record sheet, bring it to the RSO table with your model. The same admonition holds during a contest; bring your Contest Flight Card with you on the first flight; on subsequent flights, pick it up at the Return Table.

The RSO, upon clearing your model for flight, permits you to proceed to your launch area with your model.

Approach your launch area without cutting across another launch area or alley.

Place your model on your launch pad, hook it up, give it a final check, step back to your launch controller, and raise your hand. This signal alerts the RCO, who keeps a continual watch in the rangehead area for people signaling that they're ready by raising their hands. Normally, an RCO will work the launch areas in rotation, starting at Number One and proceeding down the line, taking modelers as they're ready, and finally returning again to Number One to sweep the rangehead area again. This is the only fair way to operate when a lot of people want to fly because it's *impossible* for the RCO to keep track of who's next by the order in which they raise their hands, especially if the rangehead is full of people waiting to launch.

At the very most, you'll have to wait only a few minutes until the RCO gets around to your launch area.

When your turn comes, the RCO gets verbal safety clearance from the RSO. Once the RSO has given clearance, you may insert your safety key into your launch controller. The RCO then goes into the terminal countdown over the PA system. When he gets to "zero" or "launch" or "start" (this last is the international word used to launch a model rocket), push your ignition button and launch your own model. It's then up to you to recover and return your own model.

If you have a misfire, you don't have to take your model out of the rangehead area, fix it, and run it through safety checking again. Keep some spare igniters in your pocket or at your launch controller. When the RCO calls a misfire, *get your finger off the ignition button* and remove your safety key. You may then clear your own misfire while the RCO goes on to take care of others who are ready to fly.

If somebody launches without a safety clearance and countdown, the RSO declares him out of action until the modeler finds the trouble and reports it to the RSO, who then lifts the restriction on that modeler or launch area. Yes, there are occasional accidental launches because somebody has a faulty ignition system. But, because of the PA system, the RSO, the RCO, and the separation between the launch areas, I've never seen one of these accidental launches become hazardous in any way since the Alley system was put into service in 1965. An accidental or premature launch causes so much embarassment to a model rocketeer that he quickly gets his equipment into safe shape. After all, he can't blame anybody but himself!

A recording of the flight operations on a range might sound something like this:

RCO: "Okay, the next bird to go is from Launch Area Number Six. It's a single-staged model with a Type B motor and painted fluorescent orange overall. Stations report."
Tracking East: "East go."
Tracking West: "West go."
RSO: "Safety go."
RCO: "Range is go! Safety is go! Time is running at T-minus five . . . four . . . three . . . two . . . one . . . *start!*"

(The word "start" is a universal term used internationally to indicate the instant of ignition. It means the same thing in English, French, German, Dutch, Czech, Slovak, Polish, Rumanian, Bulgarian, Italian, Spanish, Serbian, Greek, Egyptian, Hungarian, and Russian. The word "fire" must *never* be used on a model rocket range unless there is a fire in progress or a grass fire has started. At the call, "Fire!" coming from *anywhere*, everyone stops what he's doing and helps put it out immediately.)

RCO: "Model coming up on peak. *Mark!* (The word "mark" indicates the moment of maximum altitude as seen by the RCO; both trackers should stop tracking at that point, whether or not it appears to them that the model has reached peak.) Recovery system is deployed, and the descent looks good. Trackers, angles, please."
Tracking East: "Range Control, this is Tracking East. Azimuth, three-two degrees."
RCO: "Roger, East, Azimuth, three-two degrees." (The angles are always reported this way, and the RCO repeats them back to make sure he's heard

and recorded the angles correctly. Azimuth is always reported first.)

Tracking East: "Elevation, two-seven degrees."

RCO: "Roger. Elevation, two-seven degrees."

Tracking West: "Range Control, this is Tracking West. Azimuth, three-five degrees."

RCO: "Roger, West. Azimuth, three-five degrees."

Tracking West: "Elevation, three-zero degrees."

RCO: "Say again elevation. You were garbled."

Tracking West: "Elevation, three-zero degrees."

RCO: "Still garbled. Elevation, three-seven degrees."

Tracking West: "Negative, Range Control! Tracking West elevation is three-zero degrees."

RCO: "Sorry. Got it that time. Elevation, three-zero degrees. Okay, the next bird to go is . . ."

And so it goes, all afternoon. On a well-run range as many as fifty models can be launched and tracked every hour from a twelve-area rangehead.

If, during the preflight or terminal countdown, something goes wrong, *anybody* on the range may yell, "HOLD!" When the RSO and the RCO hear this, they freeze the countdown right there, recycle the count back to the range safety clearance point, and find out what the reason for the hold is. This is an additional safety measure that makes *anyone* and *everyone* on the flying field a "deputy range safety officer" who can call a halt if he sees something wrong that the RSO or RCO or anyone else doesn't. Naturally, this could get out of hand without good range discipline. Therefore, the RSO must *never* permit horseplay or false alarms on a model rocket range. It's the RSO's reponsibility to maintain order and discipline, and the RSO has the authority to have a person removed from the flying field for continued infractions of discipline.

Because of the nature of model rocketry and in spite of all the safety procedures and checks, some flights aren't successful. But, because of good range safety discipline, I've never seen any serious hazard on a model rocket range. Always keep in mind, however, that you're in a big open area with objects that are flying freely through the air at high speeds.

Therefore, I pass along to you the following tips from my own experience in watching a quarter of a million model rocket flights:

1. If you don't have range duties that require your attention, stop what you're doing during any terminal countdown and watch the model. Keep your eyes on it until you know the flight is going well or where it's going to come down.

2. Get on your feet and stay on your feet. Don't lie down and take a nap

Figure 17-8 Finally, don't forget that you have to recover your own model. A loyal and trustworthy recovery crew can be a model rocketeer's best friend.

in the warm sun. You may have to move quickly to get out of the way of an errant model. It doesn't happen often, but when it does, be prepared.

3. If a model gets into trouble in the air, don't panic and start to run. Stand still and keep your eyes on it. If it comes in your direction, you'll have to move less than 12 inches to one side to get out of its way.

4. Don't engage in horseplay or practical jokes on the range at any time, and don't let others do it, either.
5. Don't stand around in a crowd. During a heads-up flight, stand at least one arm's length away from everyone else.
6. If a model falls to the ground before its ejection charge has gone off, stand clear of it until this occurs. Then you may pick it up if it's yours.
7. When recovering a model, don't run up on it. You may trip and step on it. This doesn't help the model very much. Don't recover somebody else's model unless you're asked to do so or unless you happen to find it out in the boondocks while you're looking for your own. If you're recovering your own model, and if the competition rules permit, it's okay to snag it out of the air before it touches the ground, but don't do this with somebody else's model unless he says it's okay in advance. Best advice: let a model land on the ground, then pick it up carefully.
8. Keep a clean model rocket range. Provide trash cans, and be tough on people who don't use them.
9. Help the RSO and RCO keep spectators under control. Don't let them wander into the rangehead area or into the downrange recovery area. You and your group are responsible for the safety of everyone on the flying field. Your club will be a better one because of good range discipline, which need not be severe or unmannerly, and the spectators will respect your group because of it.
10. Keep dogs and small children under control at all times.
11. Don't try to fly at all on days with high winds and bone-dry grass on the field.
12. Use common sense, keep your cool, and have fun.

I've been on many model rocket ranges all over the United States, Canada, and Europe (East and West). It's always a pleasure to be on a well-run range. I've walked off poorly run ranges. But it doesn't take much to get a poorly run range back in the groove again. A smoothly running and efficient model rocket range where everybody takes turns at everything and everybody gets the chance to fly as much as time allows—well, *any* club or group can be very proud of that.

18

Clubs and Contests

There's a saying among model rocketeers that when two model rocketeers get together, they form a club.

Actually, it takes more than two people to have a going model rocket club. A club has lots of advantages. If you don't belong to one, you should seriously consider forming one or joining one.

It's easiest to join a club that's already in existence. But, if there isn't one around, it isn't difficult to form your own. In fact, it may come into being spontaneously when several people meet on the same field to fly together, begin to enjoy one another's company, and decide to expand the group to include others who enjoy model rocketry, too.

How do you get a model rocket club going, how do you keep it going, and how do you make it a worthwhile activity for its members, one that they'll support by coming to meetings and flight sessions and giving their time and effort to?

Let's suppose you're a lone individual who's done some flying on your own, observing the proper safety rules. You'd like to learn more about the hobby, discover what other people are doing, and encourage others to do it the right way, the model rocketry way.

If you're a young person, the first thing you should do is find an adult who's also interested in forming and belonging to a model rocket club because he can act as your advisor. As I pointed out in the preceding chapter, there is much that an adult can do that a young person can't. Your adult advisor may turn out to be one of your parents, a science teacher, a Scout or 4-H leader, or somebody at the hobby shop who's interested in young people and/ or model rocketry.

You should also try to find a sponsoring organization. A sponsor can help you find a meeting place, a flying site, and "seed money" or financial assistance to get the club started. There are many organizations today that have model rocketry programs within their own structure or that warmly encourage model rocket club activity. The Civil Air Patrol has an active model rocket program, and many Scout and Explorer groups are deeply involved in model rocketry. The 4-H Clubs also have a model rocket program. But there are other potential candidates that may sponsor independent model

Figure 18-1 Regular club meetings bring together model rocketeers with common interests to discuss technical problems, participate in training lectures, and plan future activities.

rocket clubs. These include, but are not limited to: Lions, Kiwanis, Rotary International, the local fire department, the local police department, and the YMCA and YMHA.

It will be of immense help to you in organizing the club, obtaining a sponsor, and finding meeting and flying sites if you, your senior advisor, and all your club members are also members of the National Association of Rocketry (NAR) which is the nationwide organization of model rocketeers in the United States. You can write to the NAR at the headquarters address listed in Appendix I. The organization will send you membership information and application blanks as well as information on how to get an NAR charter for your club.

I'm proud to have been one of the founders of the NAR in 1957. It's grown from an idea into an organization receiving national and international respect. Today it has thousands of members and chartered NAR Sections all over the United States. It's affiliated with the National Aeronautic Association, America's aerospace sports club. Through this NAA affiliation, the NAR is linked directly to the Federation Aeronautique Internationale (FAI) in Paris, France. National aerospace sport clubs from 61 nations belong to the FAI, and most have their national counterpart to the NAR with whom the NAR is in contact.

The NAR does many jobs. It's a nonprofit organization, which means that it's a labor of love for the people who run it.

Figure 18-2 The National Association of Rocketry (NAR) offers the model rocketeer many benefits for his membership, and many clubs are officially chartered Sections of the NAR. Here are some NAR publications. Jacket patches came from various Sections, NARAMs, and international meets.

The most important offering of the NAR for club purposes is liability insurance to protect you, your sponsor, your club, and the owner of the land you use as a flying site. Insurance is protection. It will cover personal injury and property damage caused by a model rocket if there's an accident. Having insurance doesn't mean that model rocketry's dangerous. Quite the contrary; if it were dangerous, nobody would offer to insure model rocketeers at all! As it is, it's proven itself safe enough that NAR members can be insured sight-unseen by a blanket liability policy issued by a major domestic insurance company.

Having this liability insurance will be a tremendous help in obtaining the support of sponsoring organizations and getting permission to use meeting rooms and flying fields. For a few dollars extra, the insurance coverage can be extended to cover the sponsoring organization and the owner of the flying site. Believe me, NAR insurance is a big factor when you're looking for support. Full details are available from NAR headquarters.

The NAR also charters local clubs as official Sections of the NAR. It's not easy to get an NAR Section Charter. Stiff requirements must be met by the club. It must have a minimum number of members, one of whom must be an adult over the age of 18. The club must petition the NAR for a charter, listing the names and NAR membership sporting license numbers of the club members. It must submit a set of operating rules called bylaws that must be checked and approved by the NAR. The activities of the group must be reported to the NAR at regular intervals.

The NAR wants its sections to be active, operating groups of serious model rocketeers. NAR Sections are not flash-in-the-pan neighborhood rocket clubs that have sprung up overnight and are likely to fold as quickly as they were formed. Sections must have organization, advisors, proper direction, and the ability to last.

Bylaws should be adopted at your earliest meetings. They're the operating rules of the club, a document that the club members turn to for guidance. A sample set of recommended NAR Section bylaws is presented in Appendix VI. Some club bylaws are simpler than this, some more complex. But any set of bylaws should contain the provisions given in the example. Note that the bylaws don't contain any special rules for the operation of the club's model rocket range or for the construction and flying of models by members; the basic NAR rules, standards, and regulations cover these operational aspects. Special club standards have been adopted by some groups, but these should be separate documents.

Note also that flight sessions aren't considered to be meetings of the club. It's impossible to conduct a business meeting on a model rocket range; everyone's too busy with his models or range duties. Business meetings of the club are very important and should be held regularly in a place where people can be comfortable and discussions can be held without distraction.

Although some financial support may come from a sponsoring organization from time to time, a club should try to become financially solvent. It should have its own membership dues to cover the costs of mailing meeting notices and other general club operating expenses. Most clubs charge nominal dues of a dollar a month or so. These dues should be over and above those dues to other organizations such as the NAR.

One of the most important aspects of club operations is good communications between officers and members. There should be a means for rapidly getting information to every member of the club. This may be done by a telephone committee or telephone tree where each member calls two or three other members once he gets the word by telephone from the member above him on the tree. Dates for meetings and flight sessions should be regularly scheduled and announced well in advance. If there's a change in plans such as a cancellation of a flight session because of weather, the telephone committee goes into action. A monthly business meeting is a must. Flight sessions should be held at least once a month and, preferably, every two weeks. Weekly flight sessions, on the other hand, don't give members enough time to build, repair, or modify their models.

A club newsletter should not be a substitute for face-to-face communications at a club meeting. A newsletter is more a means of publishing a printed record of meeting and flying schedules, upcoming contests, and other club data that should exist as hard copy. And the club newsletter should be a means of letting *other people* know what you're doing. It should be sent to all members as well as to local newspapers, radio and TV stations, public officials, and officers of sponsoring organizations.

Business meetings should be short and to the point without getting tangled up in personality clashes, ego trips, and the puzzlements of Robert's "Rules of Order." Elect a president or chairman who can keep the meeting on track and who doesn't like to hear himself talk. Get in touch with NASA to obtain films and other attractions to make your meetings sparkle with a program that's of interest to model rocketeers. If you do this, you'll have very little trouble getting your members to come to meetings.

I've formed five model rocket clubs and have been instrumental in the formation of several others. In all cases, there's always been one serious problem: teaching the newcomers. So in 1965, I embarked upon an experiment that ran for eight years and turned out to be highly successful. I formed the New Canaan (Connecticut) YMCA Space Pioneers Section of the NAR. Every new member had to meet a series of tough qualifications. Young people under the age of 14 had to have a parent or adult join with them and participate with them in the club activities. Attendance at meetings and flight sessions was mandatory unless the member was excused in advance. And membership was open only once each year in September. The reason for much of this was the requirement that every member of the club complete the club's 9-month training course in model rocketry taught by experienced club members. We followed a regular course plan, or syllabus, and used earlier editions of this book as the text. The training course was intended to "bring everyone up to the same level of ignorance and confusion that other club members enjoyed." It provided each member with the same foundation upon which to build his future activities in model rocketry. It eliminated a lot of trial-and-error activity by new members. And it provided a proving ground for many new ideas—the alley rangehead system, simple beginner's models, simple contests, simplified methods of calculating CP and estimating altitudes, and a number of things that are now commonplace in the hobby.

The training course made use of existing kits available in local hobby stores. Every two weeks, there was a 45-minute lecture about an assigned subject or chapter in this book, often with demonstrations. A flight session was held every two weeks to let the trainees try out the things they'd studied and built. There were regular assignments of outside work involving reading in this book and building models in the member's own workshop. There were never any workshop sessions where trainees brought their models to the club meeting to work on them together; our meeting time was far too valuable and was used instead to present information to the trainees and, once the lecture session was over, to the entire club because experienced members didn't have to show up until after the training lecture was over. The flight sessions were the testing periods during which the trainees proved how well they were getting along. Organized sporting competition was used to keep the work interesting and *to keep the experienced members on their toes;* often the trainees were pitted against the older members in contests—and won! Awards, trophies, medals, certificates, and merchandise prizes were given. Upon completion of the course, each trainee was given a diploma certifying him (or her) to be a "Compleat Model Rocketeere."

Members who'd gone through the course went on to do their own thing—competition, research, or just advanced sport modeling.

How successful was the training program concept of club operations? During every year of the club's existence between 1965 when it was formed and 1973 when I left it to move to Phoenix, there was at least one United States National Champion Model Rocketeer in the club. Twice, the New Canaan YMCA Space Pioneers won the coveted NAR National Championship Section Pennant. The club was never beaten in any local or regional competition. One club member became an FAI international medalist in the First World Championships for Space Models in Vrsac, Yugoslavia, in 1972.

I've also used the training course principles in two other clubs. It works. This statement is backed up by facts. Sometimes, the procedure must be slightly modified because of different local conditions—in Phoenix, Arizona, we could fly all winter and had to cut back on flying sessions during the summertime with its 110° F temperatures.

Many people have criticized the strong discipline and highly structured operation of the training course approach. But I have seen what it can accomplish with young people, and I wouldn't get involved with any model rocket club that didn't have some sort of training activity *because that's what a model rocket club is really good for!*

If you try the training course approach, don't try to take shortcuts by making attendance voluntary, for example. Get a commitment and require that people follow through.

Out of the training course experience, I helped develop the NAR's NARTREK (NAR Training of Rocketeers for Experience and Knowledge) program which consists of a series of different levels—bronze, silver, gold, and advanced—with standards to be met at each level. I highly recommend the NARTREK program for clubs because it solves the problem of restricting new members to admission only once each year at the start of the organized training course. Using NARTREK, each member proceeds at his own pace. This book can be of great help in proceeding through NARTREK because the program was based on this book, which in turn is based on what I learned in the New Canaan YMCA Space Pioneers training program. Your club meetings can be used to answer questions or to cover specific aspects of the NARTREK program that members may be having problems with. When all

Figure 18-3 A regular club training course was developed around earlier editions of this book and developed into the NARTREK program of the NAR. For eight years, the author taught most of this training course.

club members have won the coveted NARTREK Gold badge, you've got an experienced club! Ask NAR Headquarters about NARTREK.

Other than regular flying sessions, your club will probably be called upon to give flight demonstrations (or "demos" for short) from time to time. These may be held simply to amaze and impress your friends, school teachers, or the general public. Or they may have a more important function such as getting new members or showing local public safety officials what model rocketry is all about. Most model rocket manufacturers will be glad to help you out on a big demo, and some have special demo programs. Write to them about this.

When you put on a flight demo, do so without an admission charge, even though your club treasury could stand some bolstering. Pass the hat for donations instead, and you'll eliminate all sorts of hassles about entertainment taxes and such things.

A demo is the time to show off your club shirts, if you have them. It's a good idea to have club shirts because, as your club grows in membership, it will allow you to tell a member from a spectator. This isn't funny; such confusion has happened. Club shirts also identify club members to the RCO when they're in the downrange recovery area. And they tell spectators who's a member in case they have any questions. A plain white shirt with an NAR cloth patch sewn on (and perhaps also your club's own cloth patch which a member has designed and which you can have made at reasonable costs) and perhaps a name tag are all that are required. Club shirts (or jackets in colder climes) are better than club hats or other items of apparel because they can make your club look sharp, organized, and on the ball rather than a bunch of wildly attired characters running around in a field.

It's very important to maintain strict range discipline during a demo.

It's also important to fly only models of very high reliability. A demo is no place to try out an experimental design for the first time. Safety checking should be extra strict.

A short, successful demo is best. Start with two low-altitude models, one with a streamer, the other with a parachute. Follow this with a cloud- buster, a high-performance job with a Type C or Type D motor in it. Fly a B/G, a two-staged model, and perhaps a large model if the demo field or area is big enough. Payload models are always impressive and show that there's more to model rocketry than up-and-down, so fly a camera. Fly an egg, squeezing the flight for all the drama and suspense possible. Save your best stuff until last, leaving the spectators with something good to talk about. *Don't* fly salvos where two or more models are launced simultaneously; that's fireworks-type stuff. Have the PA system going at all times with chatter, telling people over and over again what's going on. Use the full countdown ritual and take every opportunity to show your spectators that model rocketry is not a bunch of kids playing with toys, but a serious technical hobby, a technology in miniature, the Space Age brought to Main Street USA.

Remember: more people have seen live model rockets launched than have ever watched a live launch in person from the Cape. What they see at your demos and flight sessions is, to them, the space program in miniature. Your model rockets may be the *only* rocket vehicles they've ever seen!

Contests provide lots of fun in model rocketry.

Competition began when one model rocketeer said to another, "My model rocket will go higher than your model rocket!" And the other rocketeer retorted, "Prove it!"

To keep your club's first contest from becoming too hectic and confused, schedule only two simple events such as spot landing and parachute duration. Very little special equipment is required for these events—a 100-foot tape

Figure 18-4 Contest flying within a set of standard rules allows modelers to compare their skills and rewards excellence. Scale model competitions where everyone builds the same model from identical scale data have become extremely popular.

measure and a brightly colored pole for spot landing, and two stopwatches borrowed from the school athletic department for the duration event. You should have three judges who are adults and who can render impartial decisions on all the little protests and complaints that will come up. Such minor hassles are always part of competition, even with the best aura of sportsmanship. The rules should be published or given to each competitor in advance, and you should hold a precontest briefing on the range to make sure everybody understands the rules and the operating procedures *before* the contest starts. Then, don't change the rules of procedures during the contest unless you have to, because this could change the nature of the contest and put some people at a disadvantage. If weather causes the contest to be stopped in the middle, either close the contest then and award prizes on the basis of what's happened thus far, or reschedule the contest and start the whole thing again from Square One; otherwise, different weather, wind, field, or operational conditions could give competitors who've flown an advantage over those who didn't, or vice versa.

The rules for flight duration events can be quite simple. Each flight should be timed by two timers with stopwatches in case something happens to one stopwatch. Timers start their watches when the model takes off and stop the watches when the model goes out of sight, touches the ground, or lands in a tree. Typical duration events include streamer duration, and egg loft duration. There should be a limit on the total impulse that may be used, but no limit on streamer size, parachute size, wing area, etc. This will encourage people to try different approaches to winning.

You can even conduct simple altitude events using the marker streamer method as was explained in Chapter 16.

Spot landing events are fun. There should be no limit on total impulse. Only one flight should be allowed, although a contestant can practice as much as he likes. But when he says, "This is official," that's it. Place the spot landing pole at a spot on the field selected at random, usually a spot not closer than 100 feet to the rangehead area and usually in the downrange area. Tape or staple a colored streamer to the top of it as a wind indicator, because wind is *very* important in spot landing. Most important rule: *all* recovery devices must deploy fully before the model touches the ground because you don't want people *targeting* the pole; this is a spot *landing* event. And don't let contestants tilt their launchers more than the permissible 30 degrees from the vertical; use a protractor and a weighted string to check launcher tilt. The winner is the one who lands his model with the tip of its nose closest to the pole.

If you want to run your competition under the standardized national rules, the NAR publishes the United States Model Rocket Sporting Code, a set of rules used on a nationwide basis. All NAR-sanctioned contests must be conducted according to these rules.

If your club is an NAR Section, you'll get a lot of information about running a contest from the NAR Contest & Records Committee. Having all contests run in the same fashion with the same rules on a nationwide basis makes it possible for your club to fly in somebody else's contest if it's an Open or Regional Meet (Section Meets are those sanctioned contests open only to club members). It won't take long for your club to be ready to take on another club in a sanctioned contest on another field. This can be a fun weekend for everyone. And some great club rivalries have sprung up over the years. Some NAR Regional Meets have become annual affairs—Pacific Area Regional (PAR), Washington–Maryland–Virginia (WAMARVA), and Maryland Regional (MARS), to name only a few.

Figure 18-5 Model rocketeers both young and old travel across the United States every year to the National Model Rocket Championships (NARAM) which is not only the top national meet but also an affair full of cameraderie and sportsmanship where people help each other.

Figure 18-6 Ellie Stine, then a teen-ager, won third place in the parachute duration category of the First World Championships for Space Models in Vrsac, Yugoslavia. Her father, the author, was chief of the FAI Jury. (Otakar Saffek)

Figure 18-7 The world's top model rocketeers, gold medal winners at the 1980 World Championships for Space Models held at Lakehurst, New Jersey.

Then there's the big, annual national meet, NARAM (NAR Annual Meet) where hundreds of model rocketeers from all over the United States congregate. This is the oldest continuously held national model rocket competition in the world. The first, NARAM-1, was held in Denver, Colorado, in 1959. NARAMS have since been held at the United States Air Force Academy, NASA Wallops Flight Center, the United States Army Aberdeen Proving Ground, the NASA Johnson Space Center, and several other locations.

Beyond the NARAM are the international meets and the World Championships for Space Models held every two years under the auspices of the FAI. There's no higher level of competition than the World Championships, the Olympics of model rocketry. To become a World Champion is to be accorded the title of *Weltmeister,* literally "world master."

International model rocketry has generated a tremendous amount of international goodwill, fellowship, and understanding. Model rocketry has taken Americans to Yugoslavia, Czechoslovakia, and Bulgaria; it has brought people from other nations to the United States. I have many close friends in many countries around the world, and so do other American model rocketeers. We communicate in the international language of model rocketry: bad English. Model rocketeers from the United States, Canada, Great Britain, Australia, Spain, Czechoslovakia, Poland, East Germany, West Germany,

Figure 18-8 Model rocketry, gateway to the stars . . . (Estes Industries, Inc.)

Yugoslavia, Bulgaria, Rumania, Egypt, and the USSR have competed in the World Championships which have been held since 1972. More nations will become involved as the years go by and "space modeling" spreads to more countries.

In addition to competition, there are also national and international model rocket performance records. The national certificates are given by the NAR on behalf of the National Aeronautic Association for model rocket performances that are carefully documented, checked, and certified as being the best in their respective categories in the United States. It's even more difficult to set an international model rocket record, but its reward is a huge FAI diploma in French with ribbons and wax seals on it. There aren't many of these around. To get one, you have to build and fly the best model rocket in the world.

In club work, competition, and record-setting lie the true fun and advancement of the sport of model rocketry. Clubs give you the opportunity to learn how to work with people, something that in the long run may be far more important than knowing only about rockets and other hardware. Contests run under accepted rules and procedures separate the good modelers from the not-so-good modelers and reward them accordingly. Competition sharpens your abilities, capabilities, and mental processes. Competition improves things, but always remember the Sportsman's Code:

> Keep the rules.
> Keep faith with your friends.
> Keep your temper.
> Keep a stout heart in defeat.
> Keep your pride under in victory.
> Play the game.

Epilogue

By now, it should be apparent that there's a great deal to this technological recreation called model rocketry. As many people have discovered, it involves nearly all aspects of human endeavor, just like its full-sized counterpart in astronautics.

As you progress in model rocketry or anything else you do, teach somebody else in the process. Communicate. Write it down. If I've helped you with this book, and if others have helped you, the only way you can repay us is to do the same for somebody else. Pass it on. That's been the basis of civilization for a hundred centuries.

How far do you want to go in model rocketry? To the NARAM? To the World Championships? To the moon? To the stars? It's all up to you from now on.

Bibliography

Abbott, Ira H., and Von Doenhoff, Albert E., *Theory of Wing Sections*. New York, Dover Publications, 1959.

American Radio Relay League, Inc., *The Radio Amateur's Handbook*. West Hartford, Connecticut, current edition (revised annually).

Barrowman, James S., *Calculating the Center of Pressure of a Model Rocket*, Technical Information Report TIR-33. Phoenix, Arizona, Centuri Engineering Company, 1968.

——————, *Stability of a Model Rocket*, Technical Information Report TIR-30. Phoenix, Arizona, Centuri Engineering Company, 1968.

Gregorek, Gerald M., *Aerodynamic Drag of Model Rockets*, Estes Model Rocket Technical Report TR-11. Penrose, Colorado, Estes Industries, Inc.

Hoerner, Sighard F., *Fluid Dynamic Drag*. 148 Busteeds Drive, Midland Park, New Jersey, Dr. Sighard F. Hoerner, 1958.

Hurt, H. H., Jr., *Aerodynamics for Naval Aviators*, NavAir 00-80T-80, Washington, D.C. 20025, Government Printing Office, 1965.

Lowry, Peter, and Griffith, Field, *Model Rocketry, Hobby of Tomorrow*. Garden City, New York, Doubleday and Company, Inc., 1971.

Mandell, Gordon K. *et al.*, *Topics in Advanced Model Rocketry*. Cambridge, Massachusetts, the MIT Press, 1973. (Available from the NAR.)

Pratt, Douglas R., *Basics of Model Rocketry*. Milwaukee, Wisconsin, Kalmbach Publishing Company, 1981.

Stine, G. Harry, *The New Model Rocketry Manual*. New York, New York, Arco Publishing Company, Inc., 1977.

Zaic, Frank, *Circular Airflow and Model Aircraft*. Northridge, California, Model Aeronautical Publications, 1964.

NOTE: More than 25 detailed technical reports, booklets, and other model rocket data for modelers, teachers, camp directors, and other interested individuals are available from Estes Industries, Inc. Other publications are available from Centuri Engineering Company. The National Association of Rocketry has a large number of technical reports and other publications available to its members through NAR Technical Services (NARTS). The amount of information available to model rocketeers is probably more than in any other hobby, and it's impossible to list it all because new reports and publications are continually becoming available.

Appendix I
Important Addresses

National Association of Rocketry, 182 Madison Drive, Elizabeth, PA
   15037.
National Aeronautics and Space Administration, Washington, DC 20546.
Centuri Engineering Company, P.O. Box 350, Penrose, CO 81240
Competition Model Rockets, P.O. Box 7022, Alexandria, VA 22307.
Crown Rocket Technology, P.O. Box 341, Mountlake Terrace, WA
   98043.
Estes Industries, Inc., P.O. Box 227, Penrose, CO 81240
Flight Systems, Inc., 9300 East 68th Street, Raytown, MO 64133.

Appendix II

Calculating Model Rocket Flight Performance Using the Acceleration Method

Assume the following hypothetical model rocket characteristics for this example:

Loaded weight at lift-off: 1.5 ounces = 0.094 pound
Weight at burnout: 1.3 ounces = 0.081 pound
Average thrust of motor: 0.9 pound (4 newtons)
Thrust duration of motor: 1.2 seconds

We compute the lift-off acceleration by using Equation (1) in modified form as follows:

$$A_0 = \left(\frac{F}{W_0} - 1\right) 32.2 = \left(\frac{0.9}{.094} - 1\right) 32.2 = 276.1 \text{ ft/sec/sec}$$

This is the lift-off acceleration. Since the model rocket motor consumes propellant during burning, the model gets lighter as it goes up. Its mass therefore changes, and the acceleration increases. We must therefore compute the burnout acceleration:

$$A_1 = \left(\frac{0.9}{0.081} - 1\right) 32.2 = 325.6 \text{ ft/sec/sec}$$

The model has undergone a change of acceleration, or a surge, during powered flight. To account for the surge in the flight equations requires more than simple algebra, so we must work out a method of simple

335

computing. To do this, we have to compute the average acceleration during powered flight:

$$A_{av} \;=\; \frac{A_0 \,+\, A_1}{2} \;=\; \frac{276.1 \,+\, 325.6}{2} \;=\; \frac{601.7}{2} \;=\; 300.85 \text{ ft/sec/sec}$$

The maximum velocity at burnout can then be computed using Equation (4) where $v_1 = 0$ since the model is starting from rest on the launch pad:

$$V_{max} \;=\; A_{av}t \;=\; 300.85 \,\times\, 1.2 \;=\; 361 \text{ ft/sec}$$

The average velocity during powered flight is computed using Equation (3):

$$V_{av} \;=\; \frac{V_{max}}{2} \;=\; \frac{361}{2} \;=\; 180.5 \text{ ft/sec}$$

Using Equation (2), we can then compute the burnout altitude:

$$S_{bo} \;=\; V_{av}t_b \;=\; 180.5 \,\times\, 1.2 \;=\; 216.6 \text{ feet}$$

From this point on, the flight performance calculations follow the identical method shown in the text for computing altitude gained during coasting flight, and total altitude achieved.

Appendix III

Model Rocket *CP* Calculation

From James S. and Judith A. Barrowman

Nose:

L_N = length of nose

For Cone	For Ogive
$(C_N)_N = 2$	$(C_N)_N = 2$
$X_N = 0.666L_N$	$X_N = 0.466L_N$

Conical Transition: (for both increasing and decreasing diameters)

d_F = diameter of front of transition
d_R = diameter of rear of transition
L_T = length of transition piece (distance from d_F to d_R)
X_P = distance from tip of nose to front of transition
d = diameter of base of nose

$$(C_N)_T = 2\left[\left(\frac{d_R}{d}\right)^2 - \left(\frac{d_F}{d}\right)^2\right]$$

NOTE: $(C_N)_T$ will be negative for conical boat tail.

$$X_T = X_P + \frac{L_T}{3}\left[1 + \frac{1 - \dfrac{d_F}{d_R}}{1 - \left(\dfrac{d_F}{d_R}\right)^2}\right]$$

Fins: (for multistaged models, calculate each set of fins separately, using a different X_B)

C_R = fin root chord
C_T = fin tip chord
S = fin semispan
L_F = length of fin mid-chord line
R = radius of body rear end

X_R = distance between fin root leading edge and fin tip leading edge parallel to body
X_B = distance from nose tip to fin root chord leading edge

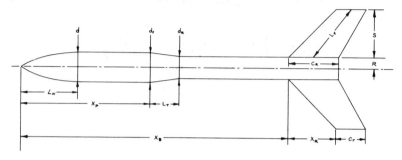

For 3 fins:

$$(C_N)_F = \left[1 + \frac{R}{S + R}\right]\left[\frac{12\left(\frac{S}{d}\right)^2}{1 + \sqrt{1 + \left(\frac{2L_F}{C_R + C_T}\right)^2}}\right]$$

For 4 fins:

$$(C_N)_F = \left[1 + \frac{R}{S + R}\right]\left[\frac{16\left(\frac{S}{d}\right)^2}{1 + \sqrt{1 + \left(\frac{2L_F}{C_R + C_T}\right)^2}}\right]$$

$$\bar{X}_F = X_B + \frac{X_R}{3}\frac{(C_R + 2C_T)}{(C_R + C_T)} + \frac{1}{6}\left[(C_R + C_T) - \frac{(C_R C_T)}{(C_R + C_T)}\right]$$

Total Values:

$$(C_N)_R = (C_N)_N + (C_N)_T + (C_N)_F + \ldots$$

(the sum of the force coefficient C_N of each part calculated)

CP Distance from Nose Tip = $\bar{X}$

$$= \frac{(C_N)_N \bar{X}_N + (C_N)_T \bar{X}_T + (C_N)_F \bar{X}_F}{(C_N)_R}$$

(the sum of the products of the force coefficient C_N and the part CP of each part divided by the total rocket C_N)

Appendix IV

Computer Program: Rocket Altitude Simulation Program "RASP-79E"

NOTE: The following program is written in BASIC language. Your particular version of BASIC may differ slightly in the commands; check your computer's instruction manual.

```
LIST
10 REM Program Code "RASP-79E"
20 REM Rocket Altitude Simulation Program
30 REM by G. Harry Stine
40 REM Version #2, 1979, English units
50 T=0
60 V=0
70 X=0
80 T1=.1
90 R=1.205
100 PRINT "Rocket gross weight in ounces:";
110 INPUT W
120 M=W*.028349523125
130 PRINT "Maximum body diameter in inches:";
140 INPUT D
150 D=D*.0254
160 PRINT "Drag coefficient:";
170 INPUT C1
180 PRINT "Motor Code:";
190 INPUT M$
200 C=R*3.14159*C1*D*D/8/M
210 PRINT "Time","Altitude","Velocity","Acceleration","Weight"
220 REM Altitude loop starts here
230 T=T+T1
240 GOTO 1000
250 A=F/M-9.8-C*V*V
260 V=V+A*T1
270 X=X+V*T1
280 IF V<0 THEN 350
290 X9=X*3.28084
300 V9=V*3.28084
310 A9=A*3.28084
320 M9=M*35.27396
330 PRINT T,X9,V9,A9,M9
```

```
340 GOTO 220
350 PRINT "Maximum altitude attained=";X9;"feet or";X;"meters."
360 PRINT "Time to peak altitude=";T;"seconds."
370 END
1000 REM Motor selection subroutine
1010 IF M$="A8-3" OR M$="A8-5" GOTO 3000
1020 IF M$="B4-2" OR M$="B4-4" OR M$="B4-6" GOTO 3200
1030 IF M$="B6-2" OR M$="B6-4" OR M$="B6-6" GOTO 3400
1040 IF M$="B14-5" OR M$="B14-7" THEN 3600
1050 IF M$="C6-3" OR M$="C6-5" OR M$="C6-7" THEN 3800
1060 IF M$="D12-3" OR M$="D12-5" OR M$="D12-7" THEN 4100
1070 IF M$="E20-4" OR M$="E20-7" OR M$="E20-10" GOTO 4400
1080 IF M$="C5-3S" GOTO 4700
1090 IF M$="F100" GOTO 5000
1100 IF M$="F7" GOTO 5200
3000 REM TYPE A8 Motor thrust routine
3010 F=0
3020 T2=.32
3030 M2=.00312
3040 IF T>T2 THEN 3100
3050 IF T=.1 THEN F=6
3060 IF T=.2 THEN F=13
3070 IF T=.3 THEN F=2
3080 M1=(M2/T2)*T1
3090 M=M-M1
3100 GOTO 250
3200 REM TYPE B4 Motor thrust routine
3210 F=0
3220 T2=1.2
3230 M2=.00833
3240 IF T>T2 THEN 3390
3250 IF T=.1 THEN F=6
3260 IF T=.2 THEN F=13
3270 IF T=.3 THEN F=5
3280 IF T=.4 THEN F=4
3290 IF T=.5 THEN F=4
3300 IF T=.6 THEN F=4
3310 IF T=.7 THEN F=4
3320 IF T=.8 THEN F=4
3330 IF T=.9 THEN F=4
3340 IF T=1 THEN F=4
3350 IF T=1.1 THEN F=4
3360 IF T=1.2 THEN F=0
3370 M1=(M2/T2)*T1
3380 M=M-M1
3390 GOTO 250
3400 REM TYPE B6 Motor Thrust routine
3410 F=0
3420 T2=.83
3430 M2=.00624
3440 IF T>T2 THEN 3560
3450 IF T=.1 THEN F=4
3460 IF T=.2 THEN F=13
3470 IF T=.3 THEN F=8
3480 IF T=.4 THEN F=6
3490 IF T=.5 THEN F=6
3500 IF T=.6 THEN F=6
3510 IF T=.7 THEN F=6
3520 IF T=.8 THEN F=6
3530 IF T=.9 THEN F=0
3540 M1=(M2/T2)*T1
```

```
3550 M=M-M1
3560 GOTO 250
3600 REM Type B14 motor thrust routine
3610 F=0
3620 T2=.35
3630 M2=.00624
3640 IF T>T2 THEN 3710
3650 IF T=.1 THEN F=6
3660 IF T=.2 THEN F=26
3670 IF T=.3 THEN F=14
3680 IF T=.4 THEN F=0
3690 M1=(M2/T2)*T1
3700 M=M-M1
3710 GOTO 250
3800 REM TYPE C6 Motor thrust routine
3810 F=0
3820 T2=1.7
3830 M2=.01248
3840 IF T>T2 THEN 4040
3850 IF T=.1 THEN F=4
3860 IF T=.2 THEN F=13
3870 IF T=.3 THEN F=8
3880 IF T=.4 THEN F=6
3890 IF T=.5 THEN F=6
3900 IF T=.6 THEN F=6
3910 IF T=.7 THEN F=6
3920 IF T=.8 THEN F=6
3930 IF T=.9 THEN F=6
3940 IF T=1 THEN F=6
3950 IF T=1.1 THEN F=6
3960 IF T=1.2 THEN F=6
3970 IF T=1.3 THEN F=6
3980 IF T=1.4 THEN F=6
3990 IF T=1.5 THEN F=6
4000 IF T=1.6 THEN F=6
4010 IF T=1.7 THEN F=0
4020 M1=(M2/T2)*T1
4030 M=M-M1
4040 GOTO 250
4100 REM TYPE D12 Motor Thrust routine
4110 F=0
4120 T2=1.7
4130 M2=.02493
4140 IF T>T2 THEN 4340
4150 IF T=.1 THEN F=6
4160 IF T=.2 THEN F=26
4170 IF T=.3 THEN F=31
4180 IF T=.4 THEN F=20
4190 IF T=.5 THEN F=11
4200 IF T=.6 THEN F=10
4210 IF T=.7 THEN F=10
4220 IF T=.8 THEN F=10
4230 IF T=.9 THEN F=10
4240 IF T=1 THEN F=10
4250 IF T=1.1 THEN F=10
4260 IF T=1.2 THEN F=10
4270 IF T=1.3 THEN F=10
4280 IF T=1.4 THEN F=10
4290 IF T=1.5 THEN F=10
4300 IF T=1.6 THEN F=10
4310 IF T=1.7 THEN F=0
```

```
4320 M1=(M2/T2)*T1
4330 M=M-M1
4340 GOTO 250
4400 REM ProJet Type E20 Motor thrust routine
4410 F=0
4420 T2=1.91
4430 M2=.0215
4440 IF T>T2 THEN 4670
4450 IF T=.1 THEN F=9
4460 IF T=.2 THEN F=10
4470 IF T=.3 THEN F=12
4480 IF T=.4 THEN F=14
4490 IF T=.5 THEN F=15.5
4500 IF T=.6 THEN F=17
4510 IF T=.7 THEN F=19
4520 IF T=.8 THEN F=21
4530 IF T=.9 THEN F=23
4540 IF T=1 THEN F=26
4550 IF T=1.1 THEN F=28
4560 IF T=1.2 THEN F=31
4570 IF T=1.3 THEN F=33
4580 IF T=1.4 THEN F=35.5
4590 IF T=1.5 THEN F=38.5
4600 IF T=1.6 THEN F=41.5
4610 IF T=1.7 THEN F=44
4620 IF T=1.8 THEN F=23
4630 IF T=1.9 THEN F=3
4640 IF T=2 THEN F=0
4650 M1=(M2/T2)*T1
4660 M=M-M1
4670 GOTO 250
4700 REM Centuri C5-3S "Super-C" Motor thrust routine
4710 F=0
4720 T2=2.1
4730 M2=.013
4740 IF T>T2 THEN 4840
4750 IF T=.1 THEN F=2
4760 IF T=.2 THEN F=7.5
4770 IF T=.3 THEN F=22.2
4780 IF T=.4 THEN F=7.5
4790  F T=.5 THEN F=5
4800 IF T=.6 THEN F=4.5
4810 IF T=.7 THEN F=4
4820 IF T=.8 THEN F=4
4830 IF T=.9 THEN F=4
4840 IF T=1 THEN F=4
4850 IF T=1.1 THEN F=4
4860 IF T=1.2 THEN F=4
4870 IF T=1.3 THEN F=4
4880 IF T=1.4 THEN F=4
4890 IF T=1.5 THEN F=4
4900 IF T=1.6 THEN F=4
4910 IF T=1.7 THEN F=4
4920 IF T=1.8 THEN F=4
4930 IF T=1.9 THEN F=4
4940 IF T=2 THEN F=4
4950 IF T=2.1 THEN F=0
4960 M1=(M2/T2)*T1
4970 M=M-M1
4980 GOTO 250
5000 REM FSI Type F100 Motor thrust routine
```

```
5010 F=0
5020 T2=.5
5030 M2=.05
5040 IF T>T2 THEN 5120
5050 IF T=.1 THEN F=88.3
5060 IF T=.2 THEN F=147.1
5070 IF T=.3 THEN F=156.9
5080 IF T=.4 THEN F=58.8
5090 IF T=.5 THEN F=0
5100 M1=(M2/T2)*T1
5110 M=M-M1
5120 GOTO 250
5200 REM FSI Type F7 Motor thrust routine
5210 F=0
5220 T2=9
5230 M2=.05
5240 IF T>T2 GOTO 5480
5250 IF T=.1 THEN F=6.9
5260 IF T=.2 THEN F=13.8
5270 IF T=.3 THEN F=20.6
5280 IF T=.4 THEN F=27.4
5290 IF T=.5 THEN F=34.3
5300 IF T=.6 THEN F=31.6
5310 IF T=.7 THEN F=28.8
5320 IF T=.8 THEN F=26.1
5330 IF T=.9 THEN F=23.3
5340 IF T=1 THEN F=20.6
5350 IF T=1.1 THEN F=17.9
5360 IF T=1.2 THEN F=15.1
5370 IF T=1.3 THEN F=12.4
5380 IF T=1.4 THEN F=9.6
5390 IF T=1.5 THEN F=6.9
5400 IF T>1.5 AND T<8.6 THEN F=7
5410 IF T=8.6 THEN F=5.6
5420 IF T=8.7 THEN F=4.2
5430 IF T=8.8 THEN F=2.8
5440 IF T=8.9 THEN F=1.4
5450 IF T=9 THEN F=0
5460 M1=(M2/T2)*T1
5470 M=M-M1
5480 GOTO 250
READY
```

Appendix V

Computer Program: Rocket Stability Calculations "STABCALC-1"

NOTE: The following program is written in BASIC language. Your particular version of BASIC may differ slightly in the commands; check your computer's instruction manual.

```
LIST
10 REM FILE "STABCALC-1"
20 REM ROCKET STABILITY CALCULATIONS
30 REM by G. Harry Stine
40 REM 1079
50 DIM T(99)
60 DIM F(99)
70 DIM X(99)
80 PRINT "ROCKET STABILITY CALCULATIONS"
90 PRINT "Write in rocket name or number"
100 PRINT "Circle dimensions used: mm in"
110 PRINT "Rocket Dimensions List"
120 PRINT "Nose length:";
130 INPUT X(1)
140 PRINT "Nose base diameter:";
150 INPUT D1
160 PRINT "Any transitions?";
170 INPUT A$
180 IF A$="NO" GOTO 490
190 PRINT "Transition 1 front diameter:";
200 INPUT D2
210 PRINT "Transition 1 rear diameter:";
220 INPUT D3
230 PRINT "Transition 1 length:";
240 INPUT L1
250 PRINT "Distance, nose tip to Trans.1 front:";
260 INPUT X(2)
270 PRINT "Any more transitions?";
280 INPUT B$
290 IF B$="NO" GOTO 490
```

```
300 PRINT "Transition 2 front dia:";
310 INPUT D4
320 PRINT "Transition 2 rear dia:";
330 INPUT D5
340 PRINT "Transition 2 length:";
350 INPUT L2
360 PRINT "Distance, nose tip to Trans.2 front:";
370 INPUT X(3)
380 PRINT "Any more transitions?":
390 INPUT C$
400 IF C$="NO" GOTO 490
410 PRINT "Transition 3 front dia:";
420 INPUT D6
430 PRINT "Transition 3 rear dia:";
440 INPUT D7
450 PRINT "Transition 3 length:";
460 INPUT L3
470 PRINT "Distance, nose tip to Trans.3 front:";
480 INPUT X(4)
490 PRINT "Fin 1 root chord:";
500 INPUT C1
510 PRINT "Fin 1 tip chord:";
520 INPUT C2
530 PRINT "Fin 1 semi-span:";
540 INPUT S1
550 PRINT "Fin 1 mid-chord line length:";
560 INPUT S2
570 PRINT "Number of fins, Fin 1:";
580 INPUT A1
590 PRINT "Radius of body at Fin 1:";
600 INPUT R1
610 PRINT "Fin 1 root to tip LE sweep distance:";
620 INPUT X(5)
630 PRINT "Distance, nose tip to Fin 1 root LE:";
640 INPUT X(6)
650 PRINT "Any more fins?";
660 INPUT D$
670 IF D$="NO" GOTO 1030
680 PRINT "Fin 2 root chord:";
690 INPUT C3
700 PRINT "Fin 2 tip chord:";
710 INPUT C4
720 PRINT "Fin 2 semi-span:";
730 INPUT S3
740 PRINT "Fin 2 mid-chord length:";
750 INPUT S4
760 PRINT "Number of fins, Fin 2:";
770 INPUT A2
780 PRINT "Radius of body at Fin 2:";
790 INPUT R2
800 PRINT "Fin 2 root to tip LE sweep distance:";
810 INPUT X(7)
820 PRINT "Distance, nose tip to Fin 2 root LE:";
830 INPUT X(8)
840 PRINT "Any more fins?";
850 INPUT E$
860 IF E$="NO" GOTO 1030
```

```
870 PRINT "Fin 3 root chord:";
880 INPUT C5
890 PRINT "Fin 3 tip chord:";
900 INPUT C6
910 PRINT "Fin 3 semi-span:";
920 INPUT S5
930 PRINT "Fin 3 mid-chord length:";
940 INPUT S6
950 PRINT "Number of fins, Fin 3:";
960 INPUT A3
970 PRINT "Radius of body at Fin 3:";
980 INPUT R3
990 PRINT "Fin 3 root to tip LE sweep distance:";
1000 INPUT X(9)
1010 PRINT "Distance, nose tip to Fin 3 root LE:";
1020 INPUT X(10)
1030 PRINT "Distance, nose tip to CG:";
1040 INPUT X(11)
1050 REM Calculation of nose***********************
1060 PRINT "If nose shape is ogive, enter 1; cone 2; parabola 3";
1070 INPUT U
1080 IF U=1 THEN 1110
1090 IF U=2 THEN 1130
1100 IF U=3 THEN 1150
1110 N2=X(1)*.466
1120 GOTO 1160
1130 N2=X(1)*.666
1140 GOTO 1160
1150 N2=X(1)*.5
1160 N1=2
1170 PRINT "Nose normal force:";N1
1180 PRINT "Nose CP:";N2
1190 IF A$="NO" GOTO 1600
1200 REM Calculation of Transition 1*************
1210 PRINT "Transition No.1:"
1220 T(1)=(D2/D1)*(D2/D1)
1230 T(2)=(D3/D1)*(D3/D1)
1240 T(3)=(T(2)-T(1))*2
1250 T(4)=(D2/D3)*(D2/D3)
1260 T(5)=1-(D2/D3)
1270 T(6)=1-T(4)
1280 T(7)=1+(T(5)/T(6))
1290 T(8)=(L1/3)*T(7)
1300 T(9)=X(2)+T(8)
1310 PRINT "Transition 1 normal force:";T(3)
1315 PRINT "Transition 1 CP:";T(9)
1320 IF B$="NO" GOTO 1600
1330 REM Calculation of Transition 2 ****************
1340 PRINT "Transition No. 2:"
1350 T(11)=(D4/D1)*(D4/D1)
1360 T(12)=(D5/D1)*(D5/D1)
1370 T(13)=(T(12)-T(11))*2
1380 T(14)=(D4/D5)*(D4/D5)
1390 T(15)=1-(D4/D5)
1400 T(16)=1-T(14)
1410 T(17)=1+(T(15)/T(16))
1420 T(18)=(L2/3)*T(17)
```

```
1430 T(19)=X(3)+T(18)
1440 PRINT "Transition 2 normal force:";T(13)
1450 PRINT "Transition 2 CP:";T(19)
1460 IF C$="NO" GOTO 1600
1470 REM Calculation of Transition 3 ****************
1480 PRINT "Transition No.3:"
1490 T(21)=(D6/D1)*(D6/D1)
1500 T(22)=(D7/D1)*(D7/D1)
1510 T(23)=(T(22)-T(21))*2
1520 T(24)=(D6/D7)*(D6/D7)
1530 T(25)=1-(D6/D7)
1540 T(26)=1-T(14)
1550 T(27)=1+(T(25)/T(26))
1560 T(28)=(L3/3)*T(27)
1570 T(29)=X(4)+T(28)
1580 PRINT "Transition 3 normal force:";T(23)
1590 PRINT "Transition 3 CP:";T(29)
1600 REM Calculation of Fin 1 **********************
1610 PRINT "Fin No. 1:"
1620 F(1)=(S1/D1)*(S1/D1)
1630 IF A1=3 THEN 1650
1640 IF A1=4 THEN 1670
1650 F(2)=F(1)*13.85
1660 GOTO 1680
1670 F(2)=F(1)*16
1680 F(3)=2*S2
1690 F(4)=C1+C2
1700 F(5)=(F(3)/F(4))*(F(3)/F(4))
1710 F(6)=1+F(5)
1720 F(7)=SQR(F(6))
1730 F(8)=1+F(7)
1740 F(9)=F(2)/F(8)
1750 F(10)=S1+R1
1760 F(11)=R1/F(10)
1770 F(12)=1+F(11)
1780 F(13)=F(12)*F(9)
1790 F(14)=X(5)/3
1800 F(15)=2*C2
1810 F(16)=F(15)+C1
1820 F(17)=F(16)/F(4)
1830 F(18)=F(14)*F(17)
1840 F(19)=C1*C2
1850 F(20)=F(19)/F(4)
1860 F(21)=F(4)-F(20)
1870 F(22)=F(21)/6
1880 F(23)=X(6)+F(18)+F(22)
1890 PRINT "Fin 1 normal force:";F(13)
1900 PRINT "Fin 1 CP:";F(23)
1910 IF D$="NO" GOTO 2550
1920 REM Calculation of Fin 2 *************
1930 PRINT "Fin No. 2:"
1940 F(31)=(S4/D1)*(S4/D1)
1950 IF A2=3 THEN 1970
1960 IF A2=4 THEN 1990
1970 F(32)=F(31)*13.85
1980 GOTO 2000
1990 F(32)=F(31)*16
```

```
2000 F(33)=2*S4
2010 F(34)=C3+C4
2020 F(35)=(F(33)/F(34))*(F(33)/F(34))
2030 F(36)=1+F(35)
2040 F(37)=SQR(F(36))
2050 F(38)=1+F(37)
2060 F(39)=F(32)/F(38)
2070 F(40)=S3+R2
2080 F(41)=R2/F(40)
2090 F(42)=1+F(41)
2100 F(43)=F(42)*F(39)
2110 F(44)=X(7)/3
2120 F(45)=2*C4
2130 F(46)=F(45)+C3
2140 F(47)=F(46)/F(34)
2150 F(48)=F(44)*F(47)
2160 F(49)=C3*C4
2170 F(50)=F(49)/F(34)
2180 F(51)=F(34)-F(50)
2190 F(52)=F(51)/6
2200 F(53)=X(8)+F(48)+F(52)
2210 PRINT "Fin 2 normal force:";F(43)
2220 PRINT "Fin 2 CP:";F(53)
2230 IF E$="NO" GOTO 2550
2240 REM Calculation of Fin 3" **********
2250 PRINT "Fin No. 3:"
2260 F(61)=S5/D1)*(S5/D1)
2270 IF A3=3 THEN 2290
2280 IF A3=4 THEN 2310
2290 F(62)=F(61)*13.85
2300 GOTO 2320
2310 F(62)=F(61)*16
2320 F(63)=2*S6
2330 F(64)=C5+C6
2340 F(65)=(F(63)/F(64))*(F(63)/F(64))
2350 F(66)=1+F(65)
2360 F(67)=SQR(F(66))
2370 F(68)=1+F(67)
2380 F(69)=F(62)/F(68)
2390 F(70)=S5+R3
2400 F(71)=R3+F(70)
2410 F(72)=1+F(71)
2420 F(73)=F(72)*F(69)
2430 F(74)=X(9)/3
2440 F(75)=2*S6
2450 F(76)=F(75)+C5
2460 F(77)=F(76)/F(64)
2470 F(78)=F(74)*F(77)
2480 F(79)=C5*C6
2490 F(80)=F(79)/F(64)
2500 F(81)=F(64)-F(80)
2520 F(83)=X(10)+F(78)+F(82)
2530 PRINT "Fin 3 normal force:";F(73)
2540 PRINT "Fin 3 CP:";F(83)
2550 REM Calculation of total rocket *********
2560 PRINT "Total rocket calculations"
2580 M1=N1+T(3)+T(13)+T(23)+F(13)+F(43)+F(73)
```

```
2590 PRINT "Total rocket normal force:";M1
2600 M2=N1*N2
2610 PRINT "Nose moment:";M2
2620 M3=T(3)*T(9)
2630 PRINT "Transition 1 moment:";M3
2640 M4=T(13)*T(19)
2650 PRINT "Transition 2 moment:";M4
2660 M5=T(23)*T(29)
2670 PRINT "Transition 3 moment:";M5
2680 M6=F(13)*F(23)
2690 PRINT "Fin 1 moment:";M6
2700 M7=F(43)*F(53)
2710 PRINT "Fin 2 moment:";M7
2720 M8=F(73)*F(83)
2730 PRINT "Fin 3 moment:";M8
2740 M9=M2+M3+M4+M5+M6+M7+M8
2750 PRINT "Total rocket moment:";M9
2760 X(12)=M9/M1
2770 PRINT "Rocket CP is located";X(12);"behind nose tip."
2780 PRINT "Rocket CG is located";X(11);"behind nose tip."
2790 B1=X(12)-X(11)
2800 PRINT "Rocket stability margin is";B1;"dimensional units."
2810 IF B1<0 THEN 2830
2820 IF B1>0 THEN 2850
2830 PRINT "Unstable rocket!!!"
2840 GOTO 2920
2850 B2=B1/D1
2860 PRINT "Rocket stability margin is";B2;"in calibers."
2870 IF B2<1.0 THEN 2890
2880 IF B2>1.0 THEN 2910
2890 PRINT "Questionable stability. Heads up!"
2900 GOTO 2920
2910 PRINT "Stable rocket design."
2920 END
READY
```

Appendix VI

Computer Program: Two-Station Alt-Azimuth Altitude Data Reduction "MRDR-1"

NOTE: The following program is written in BASIC language. Your particular version of BASIC may differ slightly in the commands; check your computer's instruction manual.

```
LIST
100 REM CODE "MRDR-1"
110 REM Two-Station Alt-Azimuth Altitude DR
120 REM by G. Harry Stine
130 REM Version #1 1979
140 B=400
150 PRINT "Azimuth #1:";
160 INPUT A1
170 PRINT "Elevation #1:";
180 INPUT E1
190 PRINT "Azimuth #2:";
200 INPUT A2
210 PRINT "Elevation #2:";
220 INPUT E2
230 A3=A1+A2
240 IF A3<180 THEN A4=A3
250 GOTO 270
260 IF A3>180 THEN A4=180-A3
270 V=B/SIN(A4*0.01745)
280 X1=V*TAN(E1*0.01745)*SIN(A2*0.01745)
290 X1=INT((10^0)*X1+.5)/(10^0)
300 PRINT "Altitude #1:";X1
310 X2=V*TAN(E2*0.01745)*SIN(A1*0.01745)
320 X2=INT((10^0)*X2+.5)/(10^0)
330 PRINT "Altitude #2:";X2
340 X3=(X1+X2)/2
350 X3=INT((10^0)*X3+.5)/(10^0)
360 PRINT "Average altitude in meters:";X3
370 REM 10% Calculation
380 P1=((X3-X1)/X3)*100
390 IF P1<0 THEN P1=P1*(-1)
```

```
400 PRINT "Percent error is:";P1
410 IF P1>10 GOTO 430
420 IF P1<10 GOTO 450
430 PRINT "Track Lost.";
440 GOTO 460
450 PRINT "Track Closed.";
460 END
READY
```

Appendix VII

Three-station Elevation-only Tracking System

Please refer to Figure 15-9 on page 273. These tables have been computed for three stations located in a straight line with the two end trackers located 100 meters (328.1 feet) from the middle tracker giving a total base line length of 200 meters (656.2 feet).

TABLE 1

Angle	End Tracker	Middle Tracker	Angle	End Tracker	Middle Tracker
1	3282.139	6564.279	26	4.203	8.407
2	820.034	1640.069	27	3.851	7.703
3	364.089	728.179	28	3.537	7.074
4	204.509	409.018	29	3.254	6.509
5	130.646	261.292	30	2.999	5.999
6	90.523	181.046	31	2.769	5.539
7	66.330	132.660	32	2.561	5.122
8	50.628	101.256	33	2.371	4.742
9	39.863	79.726	34	2.197	4.395
10	32.163	64.326	35	2.039	4.079
11	26.466	52.932	36	1.894	3.788
12	22.133	44.267	37	1.761	3.522
13	18.761	37.523	38	1.638	3.276
14	16.086	32.172	39	1.524	3.049
15	13.928	27.856	40	1.420	2.840
16	12.162	24.324	41	1.323	2.646
17	10.698	21.396	42	1.233	2.466
18	9.472	18.944	43	1.149	2.299
19	8.434	16.868	44	1.072	2.144
20	7.548	15.097	45	0.999	1.999
21	6.786	13.572	46	0.932	1.865
22	6.126	12.252	47	0.869	1.739
23	5.550	11.100	48	0.810	1.621
24	5.044	10.089	49	0.755	1.511
25	4.598	9.197	50	0.704	1.408

(Table 1, continued)

Angle	End Tracker	Middle Tracker	Angle	End Tracker	Middle Tracker
51	0.655	1.311	71	0.118	0.237
52	0.610	1.220	72	0.105	0.211
53	0.567	1.135	73	0.093	0.186
54	0.527	1.055	74	0.082	0.164
55	0.490	0.980	75	0.071	0.143
56	0.454	0.909	76	0.062	0.124
57	0.421	0.843	77	0.053	0.106
58	0.390	0.780	78	0.045	0.090
59	0.361	0.722	79	0.037	0.075
60	0.333	0.666	80	0.031	0.062
61	0.307	0.614	81	0.025	0.050
62	0.282	0.565	82	0.019	0.039
63	0.259	0.519	83	0.015	0.030
64	0.237	0.475	84	0.011	0.022
65	0.217	0.434	85	0.007	0.015
66	0.198	0.396	86	0.004	0.009
67	0.180	0.360	87	0.002	0.005
68	0.163	0.326	88	0.001	0.002
69	0.147	0.294	89	0.000	0.000
70	0.132	0.264	90	0.000	0.000

STEP 2

From Table 2:

Look up the "Subtract" number obtained above in the "Sum of Values" column of Table 2. "Height" number opposite "Sum of Values" number is the achieved altitude in meters.

Sum of Values: _____ Height: _____ meters.

TABLE 2

Sum of Values	Height	Sum of Values	Height	Sum of Values	Height	Sum of Values	Height
20000.000	1	165.289	11	45.351	21	20.811	31
5000.000	2	138.888	12	41.322	22	19.531	32
2222.222	3	118.343	13	37.807	23	18.365	33
1250.000	4	102.040	14	34.722	24	17.301	34
800.000	5	88.888	15	32.000	25	16.326	35
555.555	6	78.125	16	29.585	26	15.432	36
408.163	7	69.204	17	27.434	27	14.609	37
312.500	8	61.728	18	25.510	28	13.850	38
246.913	9	55.401	19	23.781	29	13.149	39
200.000	10	50.000	20	22.222	30	12.500	40

(Table 2, continued)

Sum of Values	Height	Sum of Values	Height	Sum of Values	Height	Sum of Values	Height	Sum of Values	Height
11.897	41	2.415	91	1.005	141	0.548	191	0.344	241
11.337	42	2.362	92	0.991	142	0.542	192	0.341	242
10.816	43	2.312	93	0.978	143	0.536	193	0.338	243
10.330	44	2.263	94	0.964	144	0.531	194	0.335	244
9.876	45	2.216	95	0.951	145	0.525	195	0.333	245
9.451	46	2.170	96	0.938	146	0.520	196	0.330	246
9.053	47	2.125	97	0.925	147	0.515	197	0.327	247
8.680	48	2.082	98	0.913	148	0.510	198	0.325	248
8.329	49	2.040	99	0.900	149	0.505	199	0.322	249
8.000	50	2.000	100	0.888	150	0.500	200	0.320	250
7.689	51	1.960	101	0.877	151	0.495	201	0.317	251
7.396	52	1.922	102	0.865	152	0.490	202	0.314	252
7.119	53	1.885	103	0.854	153	0.485	203	0.312	253
6.858	54	1.849	104	0.843	154	0.480	204	0.310	254
6.611	55	1.814	105	0.832	155	0.475	205	0.307	255
6.377	56	1.779	106	0.821	156	0.471	206	0.305	256
6.155	57	1.746	107	0.811	157	0.466	207	0.302	257
5.945	58	1.714	108	0.801	158	0.462	208	0.300	258
5.745	59	1.683	109	0.791	159	0.457	209	0.298	259
5.555	60	1.652	110	0.781	160	0.453	210	0.295	260
5.374	61	1.623	111	0.771	161	0.449	211	0.293	261
5.202	62	1.594	112	0.762	162	0.444	212	0.291	262
5.039	63	1.566	113	0.752	163	0.440	213	0.289	263
4.882	64	1.538	114	0.743	164	0.436	214	0.286	264
4.733	65	1.512	115	0.734	165	0.432	215	0.284	265
4.591	66	1.486	116	0.725	166	0.428	216	0.282	266
4.455	67	1.461	117	0.717	167	0.424	217	0.280	267
4.325	68	1.436	118	0.708	168	0.420	218	0.278	268
4.200	69	1.412	119	0.700	169	0.417	219	0.276	269
4.081	70	1.388	120	0.692	170	0.413	220	0.274	270
3.967	71	1.366	121	0.683	171	0.409	221	0.272	271
3.858	72	1.343	122	0.676	172	0.405	222	0.270	272
3.753	73	1.321	123	0.668	173	0.402	223	0.268	273
3.652	74	1.300	124	0.660	174	0.398	224	0.266	274
3.555	75	1.280	125	0.653	175	0.395	225	0.264	275
3.462	76	1.259	126	0.645	176	0.391	226	0.262	276
3.373	77	1.240	127	0.638	177	0.388	227	0.260	277
3.287	78	1.220	128	0.631	178	0.384	228	0.258	278
3.204	79	1.201	129	0.624	179	0.381	229	0.256	279
3.125	80	1.183	130	0.617	180	0.378	230	0.255	280
3.048	81	1.165	131	0.610	181	0.374	231	0.253	281
2.974	82	1.147	132	0.603	182	0.371	232	0.251	282
2.903	83	1.130	133	0.597	183	0.368	233	0.249	283
2.834	84	1.113	134	0.590	184	0.365	234	0.247	284
2.768	85	1.097	135	0.584	185	0.362	235	0.246	285
2.704	86	1.081	136	0.578	186	0.359	236	0.244	286
2.642	87	1.065	137	0.571	187	0.356	237	0.242	287
2.582	88	1.050	138	0.565	188	0.353	238	0.241	288
2.524	89	1.035	139	0.559	189	0.350	239	0.239	289
2.469	90	1.020	140	0.554	190	0.347	240	0.237	290

(Table 2, continued)

Sum of Values	Height	Sum of Values	Height	Sum of Values	Height	Sum of Values	Height	Sum of Values	Height
0.236	291	0.171	341	0.130	391	0.102	441	0.082	491
0.234	292	0.170	342	0.130	392	0.102	442	0.082	492
0.232	293	0.169	343	0.129	393	0.101	443	0.082	493
0.231	294	0.169	344	0.128	394	0.101	444	0.081	494
0.229	295	0.168	345	0.128	395	0.100	445	0.081	495
0.228	296	0.167	346	0.127	396	0.100	446	0.081	496
0.226	297	0.166	347	0.126	397	0.100	447	0.080	497
0.225	298	0.165	348	0.126	398	0.099	448	0.080	498
0.223	299	0.164	349	0.125	399	0.099	449	0.080	499
0.222	300	0.163	350	0.125	400	0.098	450	0.080	500
0.220	301	0.162	351	0.124	401	0.098	451	0.079	501
0.219	302	0.161	352	0.123	402	0.097	452	0.079	502
0.217	303	0.160	353	0.123	403	0.097	453	0.079	503
0.216	304	0.159	354	0.122	404	0.097	454	0.078	504
0.214	305	0.158	355	0.121	405	0.096	455	0.078	505
0.213	306	0.157	356	0.121	406	0.096	456	0.078	506
0.212	307	0.156	357	0.120	407	0.095	457	0.077	507
0.210	308	0.156	358	0.120	408	0.095	458	0.077	508
0.209	209	0.155	359	0.119	409	0.094	459	0.077	509
0.208	310	0.154	360	0.118	410	0.094	460	0.076	510
0.206	311	0.153	361	0.118	411	0.094	461	0.076	511
0.205	312	0.152	362	0.117	412	0.093	462	0.076	512
0.204	313	0.151	363	0.117	413	0.093	463	0.075	513
0.202	314	0.150	364	0.116	414	0.092	464	0.075	514
0.201	315	0.150	365	0.116	415	0.092	465	0.075	515
0.200	316	0.149	366	0.115	416	0.092	466	0.075	516
0.199	317	0.148	367	0.115	417	0.091	467	0.074	517
0.197	318	0.147	368	0.114	418	0.091	468	0.074	518
0.196	319	0.146	369	0.113	419	0.090	469	0.074	519
0.195	320	0.146	370	0.113	420	0.090	470	0.073	520
0.194	321	0.145	371	0.112	421	0.090	471	0.073	521
0.192	322	0.144	372	0.112	422	0.089	472	0.073	522
0.191	323	0.143	373	0.111	423	0.089	473	0.073	523
0.190	324	0.142	374	0.111	424	0.089	474	0.072	524
0.189	325	0.142	375	0.110	425	0.088	475	0.072	525
0.188	326	0.141	376	0.110	426	0.088	476		
0.187	327	0.140	377	0.109	427	0.087	477		
0.185	328	0.139	378	0.109	428	0.087	478		
0.184	329	0.139	379	0.108	429	0.087	479		
0.183	330	0.138	380	0.108	430	0.086	480		
0.182	331	0.137	381	0.107	431	0.086	481		
0.181	332	0.137	382	0.107	432	0.086	482		
0.180	333	0.136	383	0.106	433	0.085	483		
0.179	334	0.135	384	0.106	434	0.085	484		
0.178	335	0.134	385	0.105	435	0.085	485		
0.177	336	0.134	386	0.105	436	0.084	486		
0.176	337	0.133	387	0.104	437	0.084	487		
0.175	338	0.132	388	0.104	438	0.083	488		
1.174	339	0.132	389	0.103	439	0.083	489		
0.173	340	0.131	390	0.103	440	0.083	490		

Appendix VIII

Sample NAR Section Bylaws

These are sample bylaws. Individual Sections may wish to alter them or add to them, due to local circumstances. Please *do not* merely fill in the blanks of this sheet and forward it to NAR Headquarters for approval; make your own copies, and be sure you have enough for your members and a copy for NAR Headquarters. The purpose of these sample bylaws is to provide a guide for each group in drawing up its own bylaws. All bylaws and amendments thereto must be approved in writing by NAR Headquarters.

Article 1, Name: The name of this organization shall be the __

Section of the National Association of Rocketry.

Article 2, Purpose: It shall be the purpose of this Section to (a) aid and abet the aims and purposes of the NAR in (locale) _____, (b) to operate and maintain a model rocket range in accordance with the NAR Standards and Regulations, (c) to hold meetings for the purpose of aiding and encouraging all those interested in rocketry, and (d) to engage in other scientific, educational, or related activities as the NAR, the Section, or the Section Board of Directors may from time to time deem necessary or desirable in connection with the foregoing.

Article 3, Membership: All members of this Section shall be NAR members in good standing who reside in _____ (locale) _____.

Article 4, Dues: Dues shall be \$_____ per year, payable in advance. These Section dues are separate and distinct from national dues paid to the NAR. All dues monies shall be kept in a General Fund by the Secretary-Treasurer and shall be paid out by him only on order of the Section Board of Directors. Special assessments may be levied by a majority vote of the members present and voting at any meeting of the Section, provided notice of such intent is given in writing to each member at least five days preceding such a meeting.

Article 5, Meetings: Meetings of the Section shall be held at least _____times per year at times and places designated by the Section Board of Directors. Operation of the rocket range shall not be considered a meeting. A quorum shall consist of 50% of the membership of the Section. Meetings shall be conducted and governed by *Robert's Rules of Order, Revised.*

Article 6, Board of Directors: The Board of Directors of this Section shall consist of the three officers, one member at large, and a Senior Member of the NAR, who shall be designated by the NAR as Section Advisor.

Article 7, Officers: The officers of this Section shall consist of a President, a Vice-President, and a Secretary-Treasurer, all of whom shall be members of the Section and of the NAR.

Article 8, Elections: Elections of officers and members of the Board of Directors shall take place at the first meeting of the calendar year. All officers and members of the Board shall serve a term of one year. Vacancies in offices and on the Board shall be filled by nomination and election of a Section member to fill the unexpired term of office and shall take place at the Section meeting at which the vacancy is announced. Nominations for all elections shall be made from the floor, and the candidate having the largest number of votes shall be elected.

Article 9, Committees: There shall be three Standing Committees of the Section, plus such additional committees as the Board of Directors may from time to time deem necessary or desirable. The Standing Committees are as follows:

(a) Operations Committee shall be in charge of the Section's model rocket range, shall monitor the experimental technical activities of the Section members, and shall act as safety inspectors. The Chairman of this Committee shall be a Senior Member of the NAR in good standing and shall act as Range Safety and Control Officer under the NAR Official Standards and Regulations.

(b) The Contests and Records Committee shall be in charge of all arrangements for contests and shall monitor all national-record attempts by Section members. The Committee shall contain at least one Leader Member of the NAR.

(c) The Activities Committee shall be in charge of making all arrangements for all Section meetings, for conducting membership campaigns, and for carrying on public relations.

(d) The Section President shall be an ex-officio member of all committees.

Article 10, Amendments: These bylaws may be amended by a two-thirds vote of those Section members present and voting at any meeting of the Section, provided written notice of the pending amendment has been sent to the membership of the Section at least five days in advance of such meeting. No amendment of these bylaws shall be in force until approved by NAR Headquarters.

Adopted: _____

Approved by NAR Headquarters: _____

Index